Data Science Applications using Python & R:

Text Analytics

by

Jeffrey Strickland

Data Science Applications using Python & R:

Text Analytics

Jeffrey Strickland

ISBN 978-1-71689-644-6

Printed in the United States of America

lulu

Acknowledgements

I would like to think the faculty and students that I worked with at the Vellore Institute of Technology (VIT) in Vellore, India. VIT is the top private engineering university in India and they are kind enough to consider me one of their own. I would particularly like to acknowledge my friends Aswani Kumar Cherukuri, Professor & Dean School of Information Technology & Engineering (SITE), and Chandra Mouliswaran S., Assistant Professor (SG), SITE. Along with the students I taught there in September 2016 and 2018, they inspired this book. My love for India is not surpassed by my love for its people.

> *This is indeed India! ... The land of dreams and romance, of fabulous wealth and fabulous poverty, of splendour and rags, of palaces and hovels, of famine and pestilence, of genii and giants and Aladdin lamps, of tigers and elephants, the cobra and the jungle, the country of hundred nations and a hundred tongues, of a thousand religions and two million gods, cradle of the human race, birthplace of human speech, mother of history, grandmother of legend, great-grandmother of traditions, whose yesterday's bear date with the modering antiquities for the rest of nations-the one sole country under the sun that is endowed with an imperishable interest for alien prince and alien peasant, for lettered and ignorant, wise and fool, rich and poor, bond and free, the one land that all men desire to see, and having seen once, by even a glimpse, would not give that glimpse for the shows of all the rest of the world combined.*

— Mark Twain

TABLE OF CONTENTS

Preface

To write a single book about data science, at least as I view the discipline, would result in several volumes. I have come to view Data Science kind of like Engineering. We have all sorts of engineers: mechanical, electrical, civil, aeronautical, industrial, and so on. We still have them, but when we talk about them, we tend to use the general term "engineers" and their field as "engineering." I have thought about this for a few years, and I have concluded that we do something similar with the terms "data scientists" and "data science."

> *Data really powers everything that we do.*
>
> – Jeff Weiner, LinkedIn

Who needs it?

We live in a world where data is so prevalent and information derived from it is often so biased that all us of need some method for mining the right data to extract relevant information. This is easier said than done but we have data science to help.

In addition to “traditional data” (the kinds for which we can perform statistical analysis), social media and businesses have given us a new avenue of approach in data science. Not that it was never present before, but there may not been an abundance of it, and we didn’t have the right tools to deal with it.

Be that as it may, I performed my dissertation research by what was then equivalent to text analytics today, except I did it “by hand.” I still have four 3-inch binders filled with interview transcripts, observation notes, surveys, and essays from a case study in which I employed concept mapping. Concept mapping is a form of modern text analytics, but in those days, it was not an “easy” task to perform.

Concept Maps. A concept map is a graphical tool that visually represents relationships between concepts and ideas. Most concept maps depict ideas as boxes or circles (also called nodes), which are structured hierarchically and connected with lines or arrows (also called arcs). These lines are labeled with linking words and phrases to help explain the connections between concepts.

Text Analytics. Kaemingk, et al, (2016) refer to text analytics as a method for extracting concept from a body of text or *corpus* (a Latin word meaning body), which is a collection of written texts (Kaemingk, Rembado, Shroff, Entwistle, & Aiello, 2016, p. Online). So, with the advent of machine learning and numerous methods for performing text analytics, we can now perform text analytics with the same rigor we apply to statistical studies.

Applications Approach. In this book, I present text analytics using examples with substantial data, from bank product complaint transcript (CFPB, 2020) to Twitter threads of ISIS Fanboy and their opposition (Data Society, 2016). We are not going to analyze, a sentence and then take a leap of faith that the same principles apply for analyzing thousands of sentences. Instead, we will dive into analyzing, the latter. For instance, we will use complaint transcripts from the Consumer Complaint Data Catalog, with over 280,000 records.

Text Data Mining. Data mining is the semi-automatic or automatic process of taking large quantities of data and extracting data that is pertinent to an organization's data operations. Such mining pulls out data that is filtered for performing specific studies, like Tweets pertaining to the 2016 Presidential debates for studying campaign political behavior. We often make bold statements about politics and other emotionally charged topic, but we usually fail to use all available data, extract information from that data, and use the information to make informed decisions.

> *In God we trust. All others must bring data.*
>
> — W. Edwards Deming, statistician

Machine Learning (ML). ML is not the same thing as artificial intelligence (AI). ML is the science of getting computers to act without being explicitly programmed (the domain of AI). In fact, many researchers think that ML is the best way to make progress towards human-level AI. In data science, we use ML or the ML algorithms to mine data, explore mined data, classify outcomes, and predict future outcomes. I make a distinction here to emphasize that ML does not make use of traditional statistical or mathematics methods. ML algorithms may include artificial neural networks (ANN), random forests (RF), genetic algorithms (GA), and many other methods. ML is particularly useful in text analytics, where statistical methods are inappropriate.

> *I believe that at the end of the century the use of words and general educated opinion will have altered so much that one will be able to speak of machines thinking without expecting to be contradicted.*
>
> — Alan Turing, Computing machinery and intelligence

Data Analytics. Data analytics is a process of examining, cleaning, transforming, and modeling data with the goal of discovering useful information, informing conclusions, and providing decision support. Data analytics encompasses multiple methods and approaches within the realms of descriptive, predictive, and prescriptive analysis, while being used in different business, science, medical, psychological, and social science domains. The goals of data analytics are discovering useful information, informing conclusions, and supporting decision-making.

Once the data is cleaned, data analysts apply a variety of techniques, referred to as exploratory data analysis, to begin understanding the information contained in the data. Data exploration can result in additional data cleaning or additional data requirements, so these activities may be iterative. In descriptive analytics, unfolding the characteristics of the data with such measures as the mean and variance may help in understanding the data. Predictive analytics is concerned with forecasting future events based on the data that has been mined and explored. Prescriptive analytics is concerned with telling the story of why the data describes the present or predicts the future to help decision makers choose the best courses of action. While text analytics can be used for descriptive and predictive analysis, it opens up a world of potential for performing prescriptive analysis.

> *The goal is to turn data into information, and information into insight.*
>
> — Carly Fiorina, former CEO, Hewlett-Packard

Text Analytics. The use of text analytics can cut through the clutter of immediate uncertainty and changing conditions. It can help prevent fraud, limit risk, increase efficiency, meet business goals, and create more loyal customers. However, it is not foolproof. It is only effective if organizations know what questions to ask and how to react to the answers. If the input assumptions are invalid, the output results will not be accurate (Segal, 2019).

When used effectively, however, text analytics can help organizations make decisions based on highly analyzed facts rather than jump to under-informed conclusions based on emotion or instinct. It can help organizations better understand the level of risk and uncertainty they face. Text analytics can extract information that is not available through other mean—we can interrogate the data.

> *Torture the data, and it will confess to anything.*
>
> – Ronald Coase, winner of the Nobel Prize in Economics

About R and R-Studio

I wrote about 50% of this book using my R output in R Studio via R Markdown. Throughout the book, I have inserted "red-box" Markdown Notes like this:

Markdown Note. Markdown provides an authoring framework for data science. You can use a single R Markdown file to both

- save and execute code
- generate high quality reports that can be shared with an audience

R Markdown documents are fully reproducible and support dozens of static and dynamic output formats, including Word, HTML, PowerPoint, and more. Like the rest of R, R Markdown is free and open source.

About Python and Jupyter Notebook

Python is a programming language that lets you work more quickly and integrate your systems more effectively. It is the most popular programming language in the world, with a share of approximately 32% of users. For comparison, Java users makes up approximately 17% of all users (Staff, 2020). The chart below shows the comparison.

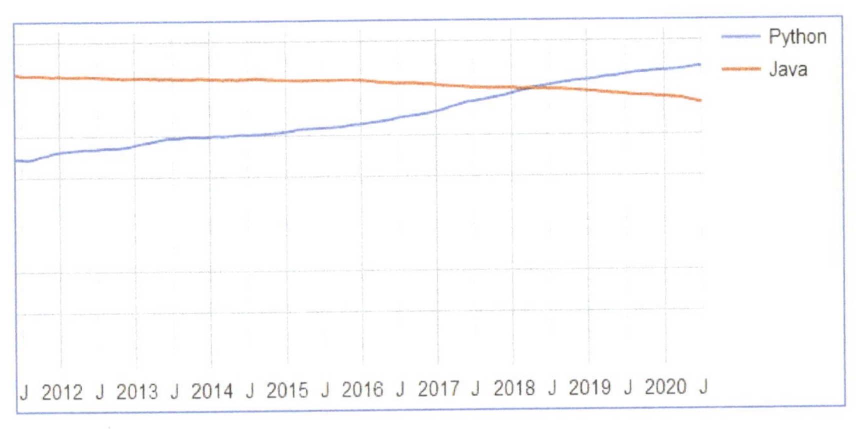

The *Jupyter Notebook* is an open-source web application that allows you to create and share documents that contain live code, equations, visualizations and narrative text (Project Jupyter, 2020). Although I will guide you through our applications that use Python, you can find a good beginner's tutorial, Google's Python Class, at:

https://developers.google.com/edu/python/

Jupyter Notebook has a "built-in" markdown feature and produces a very nice HTML output (or PDF) identical to your notebook (you can also customize the markdown).

R and Python Conventions

I wanted to make it clear when I am using R or Python, so I use the following conventions (the main body of the text is written using the Calibri sans-serif font):

- The mention of R packages within the body of the text appears as blue italicized Cambria serif font, like *readr* package.
- An R function/command within the body of the text appears as blue bold Consolas monospaced font, **`read.csv()`**.
- The mention of Python library within the body of the text appears as purple italicized Cambria serif font, like *read* package.
- A Python function/command within the body of the text appears as blue bold Consolas monospaced font, **`csv.reader()`**.

The text contains numerous chunks or snippets of code. R code and its corresponding output appear as follows:

```
R input
```

```
R output
```

Python code and its corresponding output appear as follows:

```
Python input
```

```
Python output
```

Data Access

All data sets (except for a few that are part of R packages) for this text can be downloaded from ***https://github.com/stricje1/Data***

Book Pricing

Those that follow me on LinkedIn know that my goal when I starting publishing was to minimize the cost of books, especially for parents with college student. I had to buy a book last semester for Phlebotomy—a story for another time—that was 239 pages from cover to cover (including front-matter, TOC and Index), and was not in full color, that cost $131.00. The same book, if self-published through Lulu.com (the publisher I use) the manufacturing cost would be $6.03. Lulu needs a little profit, so I could sell this book for around $10.00. Okay, let's compare: big publishing house for $131.00 or Lulu for $10.00—I might be off a penny or two, but I could attribute that to weak batteries in my calculator.

Enjoy the ride…

Related Books by the Author

Data Science Applications using R. Copyright © 2017. Jeffrey Strickland. Lulu, Inc. ISBN 978-0-359-81042-0

Predictive Crime Analysis using R. Copyright© 2018, Jeffrey Strickland. Lulu, Inc. ISBN 978-0-359-43159-5

Logistic Regression – Inside-Out. Copyright © 2017 by Jeffrey S. Strickland. Glasstree, Inc. ISBN 978-1-365-27041-3

Time Series Analysis using Open-Source Tools. Copyright© 2016, Jeffrey Strickland Glasstree, Inc. ISBN 978-1-5342-0100-2

Predictive Analytics using R. Copyright © 2016 by Jeffrey S. Strickland. Lulu.com. ISBN 978-1-312-84101-7

Data Analytics Using Open-Source Tools. Copyright © 2015 by Jeffrey Strickland. Lulu Inc. ISBN 978-1-365-21384-7

Data Science and Analytics for Ordinary People. Copyright © 2015 by Jeffrey S. Strickland. Lulu.com. ISBN 978-1-329-28062-5

Operations Research using Open-Source Tools. Copyright © 2015 by Jeffrey Strickland. Lulu Inc. ISBN 978-1-329-00404-7

Missile Flight Simulation - Surface-to-Air Missiles, 2nd Edition. Copyright © 2015 by Jeffrey S. Strickland. Lulu.com. ISBN 978-1-329-64495-3

Verification and Validation for Modeling and Simulation Copyright © 2014 by Jeffrey S. Strickland. Lulu.com. ISBN 978-1-312-74061-7

Mathematical Modeling of Warfare and Combat Phenomenon. Copyright © 2011 by Jeffrey S. Strickland. Lulu.com. ISBN 978-1-45839255-8

Simulation Conceptual Modeling. Copyright © 2011 by Jeffrey S. Strickland. Lulu.com. ISBN 978-1-105-18162-7.

Discrete Event Simulation using ExtendSim 8. Copyright © 2010 by Jeffrey S. Strickland. Lulu.com. ISBN 978-0-557-72821-3

Chapter 1 – What is NLP all About?

I am undecided about what came first, the chicken or the egg. It was like this with text analytics and natural language processing (NLP). Then I had a job where I performed text analytics and recalled that had done it before, but without the aid of software—I did by hand. So, with a computer and armed with the proper software, I began using NLP to perform text analytics. The point of this is to say that NLP is a tool used for doing text analytics, and is usually associated with software of a programming language like Python.

I also studied cognitive linguistics when I was studying educational mathematics at the University of Northern Colorado. From the cognitive linguistics perspective, I saw NLP as a subfield of linguistics. Thus, to give a definition:

Definition 1.1. *Natural Language Processing is a subfield of linguistics used to analyze a body of text (a corpus) using computer software.*

1.1 What is Natural Language Processing?

We need to start this with a different question: what is natural language? The oxford dictionary says that natural language is, "a language that has developed naturally in use (as contrasted with an artificial language or computer code)." (Lexico.com, 2020)

Might it be the case that natural language actually means "natural language," nothing more and nothing less? From the Latin, *naturalis*, the etymology of "natural", is 'birth, nature, or quality'. So, let's say that natural language is something that we are born with that sounds like noise until we learning to say "dadah" for "daddy." However, the noise was also language, from the Latin *lingua*, meaning tongue, which implies something that parents eventually come to understand, though perhaps vaguely. Even people who never learn to spell or to write have language.

So, when we say "natural language processing," does it rule out text? We can certainly write the symbolic representations of natural language on a piece of paper or type it on a computer (I would say typewriter, but fear some would have to Google it). So, is there a difference between natural language processing and text analytics.

Let's take this a step further. What does analytics mean? If I look at the entomology or "analytics," I find its root in Greek, which is not at all uncommon with English words. It is derived from *ἀnalytiká*, meaning the "science of analysis."

Woolman (2006) states: "Science is based primarily on an empirical approach to gathering information—an approach that relies on systematic observation."

Well, that left me hanging. So, what is "analysis"? It is a combination of *ana-*, meaning up, and *luen*, meaning loosen, to form, *analuen*, meaning "unloose" or "unravel" (all of these are Greek roots). So, text analytics is "the science of unloosening text, and text is from a Latin root meaning "woven." Without the further study of Latin and Greek roots, here is my definition of **text analytics**.

Definition 1.2. *Text analytics is a structured process of unraveling text to a useful semantic unit with the goal of discovering useful information, informing conclusions, and providing decision support.*

Saving us some time and some more Latin, processing means "to perform a series of operations," so NLP is structure process and we can use it as one of the processes for unraveling text. Therefore, there is a distinction between text analytics and NLP.

I would venture to say that text does not come off of the tongue, but something that comes off of the tongue can be represented as text, and we usually call the collection of those representations, transcripts, or "something that is written across (more Latin), or make a copy in writing." One problem arises here. Natural language may not provide uniform meaning. In English, this can be a perplexing problem (not so much in Greek). For this reason, we adopt a lexicon when using NLP (more on this later).

However, can I process natural language on my computer without transcribing it. I suppose some would say that we have voice recognition software to do that, but is the software not "transcribing" it to something the computer can understand?

Some say that natural language processing is different because it is understanding underlying meaning, while text analytics does not consider the semantics in the text.

So, my thought is this:

We try to understand the progression of something that from birth comes off of the tongue by unloosening that which is woven.

Or restated:

> ***Definition 1.3.*** *Natural language processing is a structured method of transforming natural language into semantic units using computers to derive useful information from it.*

Just don't ask me to represent the cries of a baby in any form.

1.2 Objections

I believe my colleagues in artificial intelligence would disagree with **Definition 1.3**. They seem to think that natural language processing is the process of analyzing text (Milward, 2020), while I have claimed that text is not necessarily natural language. Now, that is okay because I do not believe that artificial intelligence is involve in any of this and that Alan Turing's dream has not yet become reality, as much as I admire Alan Turing.

I also believe that my data science colleagues would object, for their most common definition is wrapped into text mining, which they also say is the same things as text analytics and is drawn from Oxford as, "the process or practice of examining large collections of written resources in order to generate new information." (Expert Systems Team, 2016)

Data scientist continue by saying that the goal of text mining is "to discover relevant information in text by transforming the text into data that can be used for further analysis." Wait, there is more. "Text mining," they say, "accomplishes this through the use of a variety of analysis methodologies; natural language processing (NLP) is one of them." (Expert Systems Team, 2016)

Now, you can see my own confusion in that the title of the book is "Text Analytics..." and nothing about NLP. Given that alone, it would appear that I am agreeing with other data scientists. But, all of that is well and good, but when rubber meets the road, none of the foregoing matters, because we do not care about lexical semantics. We only care about doing something in order to make a machine analyze token, like words and sentences, so that we can make inferences, predictions, and so on.

So, notwithstanding the various definitions, why is making meaning of words so important?

1.3 Who Else Cares?

There are other disciplines that use NLP. Some include computer science, artificial intelligence, and linguistics. The way each of these define NLP is based on what they use it for, so AI's meaning for NLP is good for them and in that context, I have no argument to present them.

1.4 Importance Natural Language Processing

In data analytics, NLP has become more and more useful by the day. I recall the days I worked on my dissertation, "How Students Make Meaning in a Reformed Calculus Course." It was the first mathematics dissertation on record that was based on qualitative analysis. You could say it was the first NLP discipline in mathematics. I used a software tool to analyze transcripts of conversations I had with several students who were participating in my case study. The software, 1981, QSR NUD*IST was one of the first qualitative research programs of its kind. The product stood for "Non-Numerical Unstructured Data Indexing Searching and Theorizing." It helped to change the way that qualitative research is conducted. In 1998, I was using version N4 which introduced the ability to work with statistical data and offered tools for merging and sharing information easily.

Now, at least from the perspective of a data scientist, Python seems to be the best tool (and programming language) for text analytics, followed by R. I have even written SAS programs that do a little NLP. One data scientist told me, and they were serious, that if you did not program with Python, then you were not a data scientist, which obviously is a ridiculous thought. Anyone who knows anything about any kind of "scientific" analysis knows there is a tool for every problem. That is, the problem drives the tool selection, not the opposite. If the tool we use drove the problem we were solving, then we would not solve very many problems.

On top of Python, a number of specialized "add-in" (we know them as libraries or packages) offer a wide range of excellent tools for performing NLP and text analytics. This progression from the one and almost only tool (QSR NUD*IST) to a plethora of software tools is representative of the importance that NLP and text analytics has assumed. To further see how important it has become, let's look at some real problems it is solving.

1.5 Natural Language Processing Applications

1.5.1 Compliance

In banking, NLP is being employed to detect and predict compliance (violations) of federal regulations. This is important because a not compliant bank, be fined, forced to pay damage, shut down, and/or lose their federal insurance through the Federal Deposit Insurance Corporation (FDIC).

In January 2019, USAA was forced to pay a $3.5 million civil penalty and make $12 million in restitution to about 66,000 customers to settle charges that it violated the federal banking regulations.

In 2017, Wells Fargo was involved in an ongoing crisis concerning fraudulent business practices in many areas of its banking enterprise. The outrage continues to unfold and touch different areas of Wells Fargo's operations, from unwanted credit card accounts to unauthorized auto insurance products to excessive fees for merchant banking. In April 2018, Federal regulators hit Wells Fargo with an astounding $1 billion civil penalty, and the bank pledged to repay affected customers $182 million.

1.5.2 Customer Satisfaction

The way in which we use to measure customer satisfaction was by surveys. However, to get one to take a survey often involved incentives. Moreover, there are people who thrive on taking surveys and those that all appalled by them. I think you see where this is headed: bias, and a lot of it. But when that was all we had, the that was all we had.

Social media has changed that a little. But there is still a certain type of customer that loves to complain on social media and another type that is happy but are content not to say how happy they are on social media.

So, what is remaining that could be a testbed for customer satisfaction? Phone calls? Have you ever noticed when you call the service, you get the recording that says something like, "This call may be recorded for quality assurance"? That gives the call center permission to "eavesdrop" or record the call. So, we get the transcripts for all of those calls, and that is the data we now use to measure customer satisfaction. But wait, call now and you get six menu options, which takes you to four more once you make a selection, and that leads to ad infinitum. By the time you talk to a real person, if that happens at all, you are so angry that we do not want to know how satisfied you are. To add injury to insult, the feds want everything automated!

We can only do what we can do...

1.5.3 Customer Complaints

If we revisit the regulatory compliance issue with banks, I can give you a concrete example. Go to the federal complaints website, *Consumer Complaint Database*[1], and download complaints under "Download option and API," and use the filter SCRA, which stands for The Servicemembers Civil Relief Act (Americanbar.org, 2018). You should get a little over one million records. There are other regulations imposed by the feds, for our protection, like Reg E and Reg Z.

Regulation E is a Federal Reserve regulation that outlines rules and procedures for electronic funds transfers (EFTs) and provides guidelines for issuers and sellers of electronic debit cards. The Federal Reserve issued Regulation E as an enactment of the Electronic Funds Transfers Act, a law passed by the U.S. Congress in 1978, which is a means of protecting consumers engaged in these kinds of financial transactions. (Kenton, 2019)

Regulation Z is the Federal Reserve Board regulation that implemented the Truth in Lending Act of 1968, which was part of the Consumer Credit Protection Act of that same year. The act's major goals were to provide consumers with better information about the true costs of credit and to protect them from certain misleading practices by the lending industry. Under these rules, lenders must disclose interest rates in writing, give borrowers the chance to cancel certain types of loans within a specified

---

[1]Link: *https://www.consumerfinance.gov/data-research/consumer-complaints/#download-the-data*

period, use clear language about loan and credit terms, and respond to complaints, among other provisions. (Kenton, Regulation Z, 2019)

Then there's *Regulation B* and *Regulation T*, and you can Google for the rest of them.

Of course, there are complaint hotlines and departments that handle nothing except complaints. We generally cannot use these as a measure of satisfaction, because the people who do not have complaints aren't calling. What we can do is use the transcripts and machine learning algorithms to detect and predict complaints and classify them according to the type of complaint, which could be by service or product or some other grouping.

1.5.4 Document Analysis

Let's take *Regulation U* as an example. We have not read it or seen its title. We just heard someone mention, "Reg U." Let's analyze the document using topic analysis (we'll cover the details later). Let's also get words appearing only 20 times or more (not including words like "the", "and", "but", and so on). We pull in the text from a website directly into R Studio. When we run our analysis, we will see the following words:

```
"credit", "lender", "loan", "margin", "regulation",
"stock"
```

Now, we know that Reg U is probably is a "regulation" about "loans and lenders" in the context of the "margin stocks."

Now let's get the title of Regulation U: "Credit by Banks or Persons other than Brokers or Dealers for the Purpose of Purchasing or Carrying Margin Stocks."

Our topic analysis picks topics through word frequency that tells us a lot about Reg U, now that we see the title for comparison, and we have never seen or read it.

1.5.5 Tweet Sentiment

Never underestimate the power of a Tweet. With Tweet sentiment analysis, we can see who is will really win the presidential race, and whether or not an ISIS attack is imminent. Figure 1-1 shows the sentiment in a collection of tweets from opposers of ISIS Fanboy.

When I started Tweet analysis, it was easy to download tweets from anyone, anywhere. Then it got tougher as privacy became an issue and authorization and authentication became necessary. Then it began to cost to download tweets. We still have to pay for the service, but the tools available are much better. I recommend two Twitter APIs.

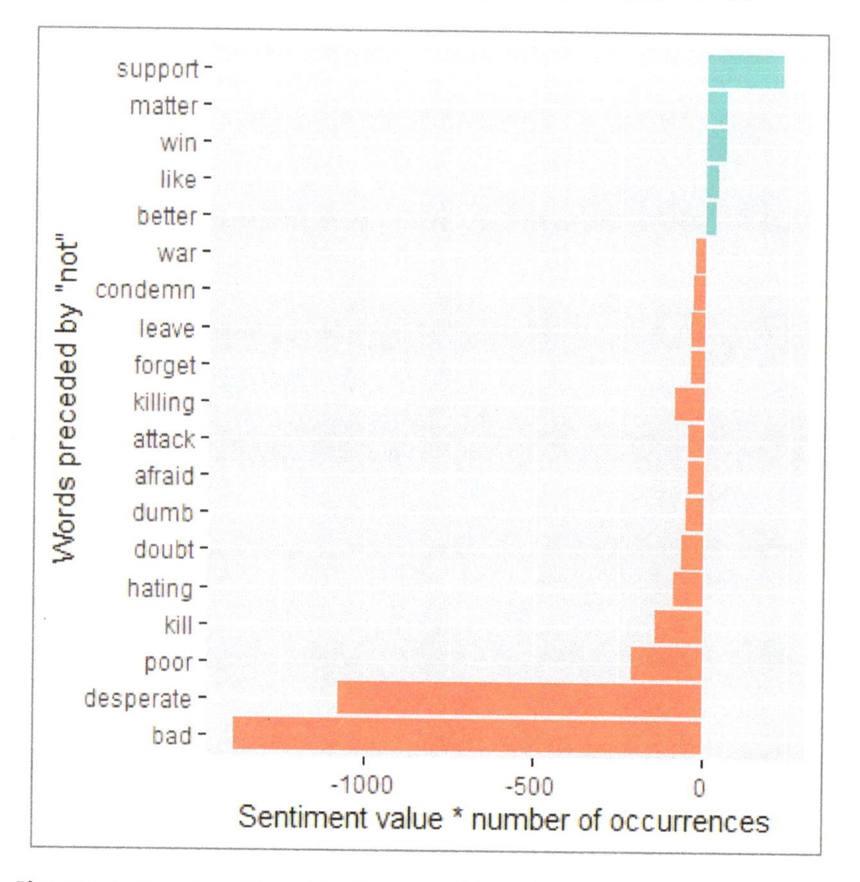

Figure 1-1. Tweet sentiment by the opposition of ISIS Fanboy. The plot shows word frequencies and magnitude (positive or negative) of word preceded by "not."

ExportTweet is a pay-as-you-go service and probably the best Tweet downloader tool available. With ExportTweet, you can download tweets of any user in a simple readable format, and it provides the following features:

- Twitter Account Analytics: Track comprehensive insights of brands, competitors and even individuals along with the latest 3200 tweets of the desired user in the Excel spreadsheet.

- Download Tweets: Download tweets from any user on Twitter. Export tweets along with various Twitter data headers in comma-separated values (CSV)/Excel format to conduct tailored offline analysis.

- Hashtag Analytics: Analyze keywords / hash tags / user mentions on Twitter in real-time. Gather and analyze the latest tweets where the searched keyword or hash tag is mentioned.

- Followers Analysis: Download followers of any public account on Twitter along with comprehensive insights report. Conduct follower analysis with crucial Twitter data exported under various headers in CSV/Excel.

It's a one-click process to download tweets. Just enter the account handle, add to cart, and download the report. It offers information like:

- Tweet day, date, time.
- Tweet likes, retweets.
- Type of Tweet (text, retweet, reply)
- Tweet source.

FollowerAnalysis is another tool like ExportTweet that helps you to download tweets of any user. When you go to the webpage of FollowerAnalysis, click on the tab "User Tweets & Analysis." Enter the Twitter handle of the user whose tweets you wish to download.

Next, you will be asked for authentication. Authenticate with your Twitter account and confirm the Twitter handle of the user. Next, pay $25 and your Tweet sheet is ready to download.

R packages

twitteR is an R package which provides access to the Twitter API. Most functionality of the API is supported, with a bias towards API calls that are more useful in data analysis as opposed to daily interaction.

Rtweet is an R package that offers much more functionality than tweetR is better for some users, like non-developers (see Table 1-1).

Table 1-1. R Twitter package comparison.

Task	rtweet	twitteR
Available on CRAN	✓	✓
Updated since 2016	✓	✗
Non-'developer' access	✓	✗
Extended tweets (280 chars)	✓	✗
Parses JSON data	✓	✓
Converts to data frames	✓	✓
Automated pagination	✓	✗
Search tweets	✓	✓
Search users	✓	✗
Stream sample	✓	✗
Stream keywords	✓	✗
Stream users	✓	✗
Get friends	✓	✓
Get timelines	✓	✓
Get mentions	✓	✓
Get favorites	✓	✓
Get trends	✓	✓
Get list members	✓	✗
Get list memberships	✓	✗
Get list statuses	✓	✗
Get list subscribers	✓	✗
Get list subscriptions	✓	✗
Get list users	✓	✗
Lookup collections	✓	✗
Lookup friendships	✓	✓
Lookup statuses	✓	✓
Lookup users	✓	✓
Get retweeters	✓	✓
Get retweets	✓	✓
Post tweets	✓	✓

Task	rtweet	twitteR
Post favorite	✓	✗
Post follow	✓	✗
Post mute	✓	✗
Premium 30 day	✓	✗
Premium full archive	✓	✗
Run package tests	✓	✗

1.6 Summary

There are many more applications of text analytics. For example, Figure 1-2 shows the most frequently occurring words in the Bible book of 1st John. This could be interpreted as the topics of the book, or its theme: *"Whoever loves the Son, the Father abides, and they are children of God, not the world, and they know life."* And Bible scholars generally take this to be the theme of 1st John.

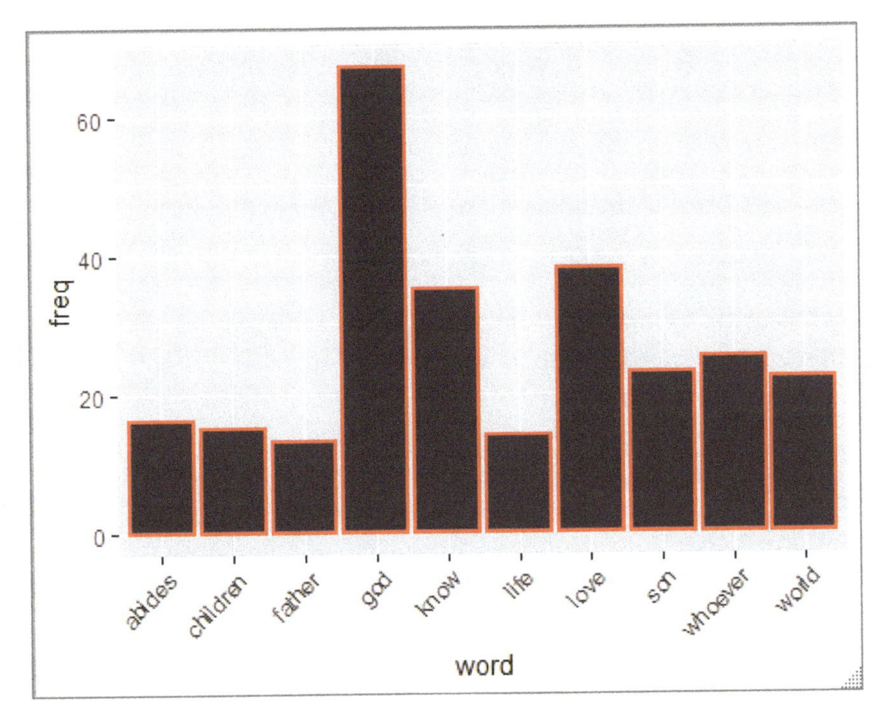

Figure 1-2. Word frequencies of the book of 1st John in the Bible.

I suppose you could convert this book to text only and get the gist of it without reading any of it. And if we wanted to know more about the important words in 1st John, where important translates to word frequency, we could plot a so-called *word cloud*. Figure 1-3 shows the word cloud for 1st John. The bigger the word, the more it occurs.

Figure 1-3. *The word cloud for 1st John.*

These are a few of the applications we will cover in this book.

1.7 Code Inputs and Outputs

As this textbook includes coding with Python and R, we will take a moment to discuss these tools and how the inputs and outputs are displayed in the book.

1.7.1 Python with Jupyter Notebooks

First, I use Anaconda's open-source Individual Edition as an easy way to perform Python data science and machine learning. Anaconda is the birthplace of Python data science. Its cloud-based repository has over 7,500 data science and machine learning packages. Packages can be easily installed using the conda-install command. Figure 1-4 shows the Anaconda Navigator console.

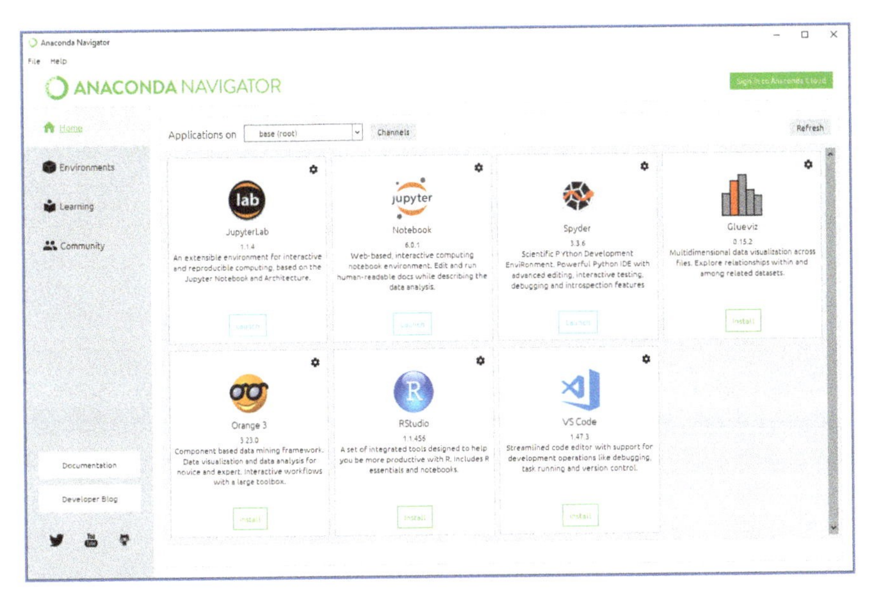

Figure 1-4. The Anaconda Navigator window with tools, including JupyterLab, Jupyter Notebook, R Studio and others.

We will be using Jupyter Notebook for Python programming and RStudio for R programming. Figure 1-5 shows a typical Jupyter Notebook with inputs (`In[n]`) and outputs (`Out[n]`).

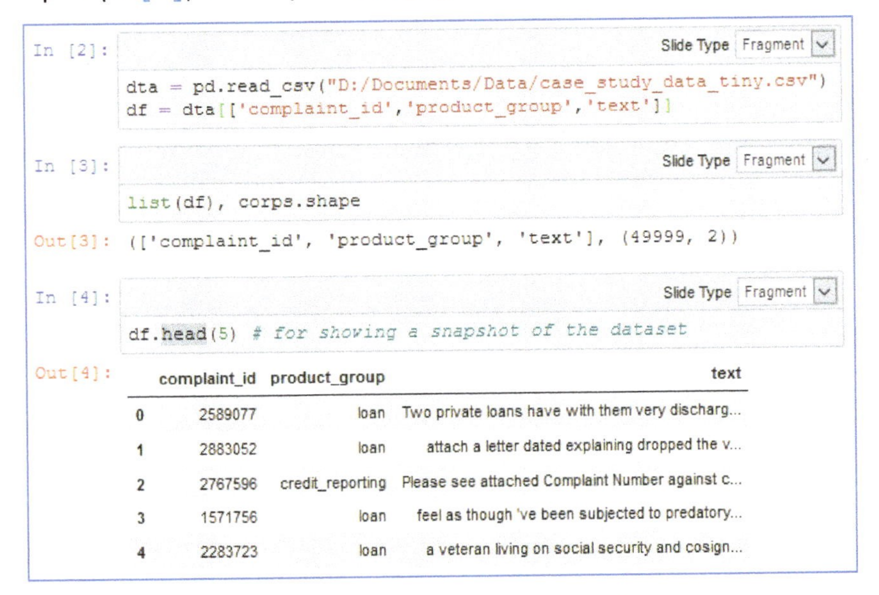

```
In [2]:                                        Slide Type  Fragment
dta = pd.read_csv("D:/Documents/Data/case_study_data_tiny.csv")
df = dta[['complaint_id','product_group','text']]

In [3]:                                        Slide Type  Fragment
list(df), corps.shape
Out[3]: (['complaint_id', 'product_group', 'text'], (49999, 2))

In [4]:                                        Slide Type  Fragment
df.head(5) # for showing a snapshot of the dataset
Out[4]:
```

	complaint_id	product_group	text
0	2589077	loan	Two private loans have with them very discharg...
1	2883052	loan	attach a letter dated explaining dropped the v...
2	2767596	credit_reporting	Please see attached Complaint Number against c...
3	1571756	loan	feel as though 've been subjected to predatory...
4	2283723	loan	a veteran living on social security and cosign...

Figure 1-5. Jupyter Notebook screenshot with input and output

In the book, we show these inputs and outputs by highlighted code boxes, corresponding to In[3] and Out[3] for this example (rather than using screenshots that could be difficult to read).

```
list(df), df.shape
```

```
(['complaint_id', 'product_group', 'text'], (268380, 3))
```

1.7.2 R with RStudio

R Studio is an integrated development environment (IDE) for R. It includes a console, syntax-highlighting editor that supports direct code execution, as well as tools for plotting, history, debugging and workspace management. It is available in open source edition and runs on the desktop (Windows, Mac, and Linux) or in a browser connected to RStudio Server. Figure 1-6 depicts the IDE window, showing the Script palette (upper left), Console (lower left), Environment pallet (upper right) and Package palette (lower right).

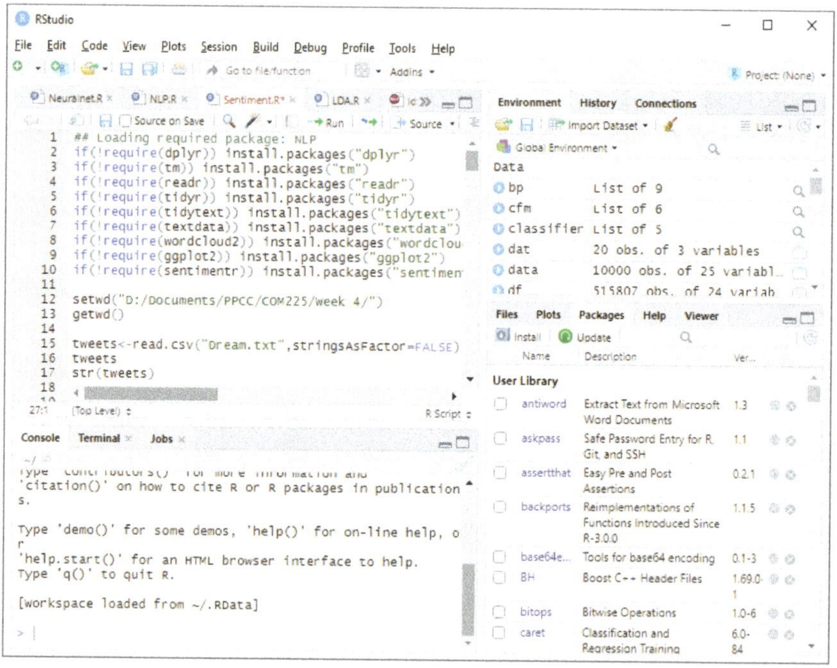

Figure 1-6. RStudio integrated development environment (IDE) for R.

In the book, we use different input and output boxes to distinguish between Python and R. Below, we show an R input box and output box, respectively.

```
df <- read.table("https://s3.amazonaws.com/assets.datacamp.com/b
log_assets/test.txt",  header = TRUE)
inspect(df)
```

```
<<SimpleCorpus>>
Metadata:  corpus specific: 1, document level (indexed): 0
Content:  documents: 20
```

1.7.3 Basic Online Tutorials

If you are unfamiliar with RStudio, Jupyter Notebook, or both, there are numerous introductory tutorials available online. Here are some of my favorites.

R Studio

R Studio Tutorial at ***http://web.cs.ucla.edu/~gulzar/rstudio/basic-tutorial.html***

R Tutorial at ***https://www.statmethods.net/r-tutorial/index.html***

Markdown Note. Markdown for R code always begins with

```{r sec2-subset, eval = TRUE, cache = TRUE}

The symbol ``` are next to the number 1 key or above the Tab key on most keyboard. The markdown will end with ```. You can also add comments in the code, just like you would normally do, starting with the # key. Normal text would be typed as you would with a text editor, like: This code will print your graph. Heading can be marked with # (level 1), ## (level 2), and so on.

Jupyter Notebook

- Notebook Examples at *https://jupyter-notebook.readthedocs.io/en/stable/examples/Notebook/examples_index.html*
- Notebook Tutorial at *https://nbviewer.jupyter.org/github/jupyter/notebook/blob/master/docs/source/examples/Notebook/Running%20Code.ipynb*
```

Markdown Note. For markdown in a Jupyter Notebook, there is a pull-down menu as shown in Figure 1-7. Notice that headings are marked like they are in R and that bullets are mark with *. There are many markdown options that are documented at:

https://jupyter-notebook.readthedocs.io/en/stable/examples/Notebook/Working%20With%20Markdown%20Cells.html

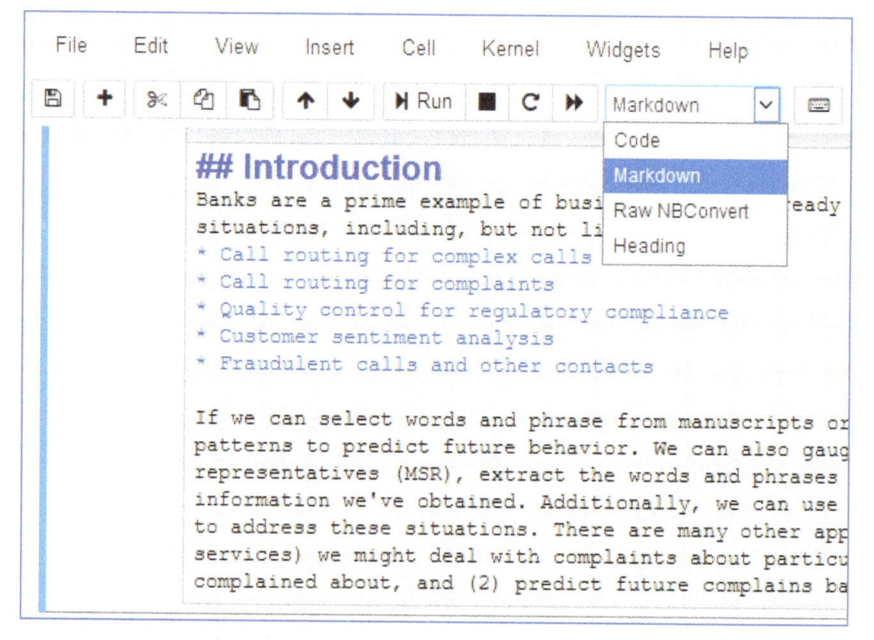

Figure 1-7. *Jupyter Notebook markdown method.*

As a general note, most of the output such as warnings, like "`Deprecated, use tibble::rownames_to_column() instea`," are omitted as the may not apply to your systems, configurations, etc. In some cases, only a few presentive bits of output are shown, like lines of text.

Chapter 2 – Introduction to Text Mining

2.1 Introduction

This is our first actual lesson on text analytics, so we will do some necessary but basic preprocessing to prepare for our analysis. This includes converting the text to lower case, removing numbers and stop-words, combining words that need to stay together (like "data science"), and putting our text into a dataframe. We will have more to say about these things below in Section 2.3.

The text we will use is a collection of texts, specifically a few blogs I have written. The complete set of blogs comprise what we call a *corpus*, which is Latin for "body" or "body of texts" in this instance. You may have heard *corpus* used in city names like Corpus Christi, which literally means "the body of Christ."

Definition 2.1. *A corpus is a collection of written texts, especially the entire works of a particular author or a body of writing on a particular subject.*

2.2 Load the R packages

We will use a "standard" set of R and Python "functions" from existing libraries or packages. Most of the time, we call (load) these functions when we need them, but here we will load them first and discuss them when we use them. For this situation, we load the following R packages for natural language processing. The manner in which we load them allows use to keep the calls in our code. If they are already installed, then the call to install them is ignored and the library of functions are loaded. Otherwise, they are installed and loaded.

```
if(!require(knitr)) install.packages("knitr")
if(!require(dplyr)) install.packages("dplyr")
if(!require(tm)) install.packages("tm")
if(!require(readr)) install.packages("readr")
if(!require(tidyr)) install.packages("tidyr")
if(!require(tidytext)) install.packages("tidytext")
if(!require(textdata)) install.packages("textdata")
if(!require(wordcloud2)) install.packages("wordcloud2")
if(!require(ggplot2)) install.packages("ggplot2")
if(!require(ggraph)) install.packages("ggraph")
```

In R-speak, we are saying that if a package has not been installed then install it and load it; otherwise load it.

2.3 Corpus Preprocessing

After we load the required packages in R, we perform a series of preprocessing steps. For now, we'll inspect some of the text I have previously loaded, and then load it for our new script after discussing data import methods.

```
inspect(docs)
```

```
<<SimpleCorpus>>
Metadata:  corpus specific: 1, document level (indexed): 0
Content:  documents: 20

A one-eyed man in the kingdom of the blind.txt

A one-eyed man in the kingdom of the blind:\Predicting the Unpre
dictable\nâ\200œAlmost nobodyâ\200\231s competent, Paul. Itâ\200
\231s enough to make you cry to see how bad most people are at t
heir jobs. If you can do a half-assed job of anything, youâ\200\
231re a one-eyed man in the kingdom of the blind.â\200\235 â\200
"Kurt Vonnegut, Player Piano\nAbstract\nThis article is about Pr
edictive Modeling. It explores the appropriateness of modeling i
n general and predictive modeling in particular, as well as exam
ining some pitfalls. Modeling is the process of formulating and 
abstracting a representation of a real problem, based on simplif
ying assumptions. Thus, no model is an exact representation of r
eality. Said a different way, a model cannot fully represent a c
omplex problem, but can provide some insight into the problem an
d assist decision makers with applying solutions.
```

As we inspect the data, using the first document as an example, it is apparent that the text is unstructured from a machine processing perspective. That is, the text contains letter, punctuation, special characters, numbers, and so on. These present irregularities and ambiguity, which makes it difficult for a machine learning algorithm to understand. For instance, in the phrase, "a one-eyed man," the word "man" is interpreted differently that the word "Man." Therefore, we must *standardize* the text before processing it with our algorithms. We often refer to this process as *cleaning the text* or *preprocessing the text*. In any case, processing is absolutely necessary in order to make meaning of it.

There are standard procedures for preprocessing text. We will use the following process:

Step 1: Load the text (data)

There are numerous ways of importing text data into R Studio or Jupyter Notebook. For instance, we can use an URL to import text files located in GitHub, or we can use a directory path on our local drive to import text files. In R Studio, some packages contain data sets that can be imported using a data statement. We generally use other processes or tools, like Hive in Hadoop or Structured Query Language (SQL) in some database or warehouse. We'll talk about the details of loading text in Section 2.4.

Step 2: Tokenize the text

A *token* is a unit of measure for text, like letters are tokens for words, words are tokens for sentences, and sentences are tokens for paragraphs.

Definition 2.2. *Tokenization is the process of splitting the text into tokens like words, sentences, or paragraphs.*

When we process the text with a machine learning algorithm, we must have a token in mind stemming from the problem we are addressing. For instance, if we plan to use the algorithm called "bag-of-words," then we word *tokenize* the text as words. Also, is we are merely wanting to derive *bigrams* (we will talk about its probabilistic meaning later), then we would tokenize the text as words.

Step 3: Convert to lower case.

Recalling that the computer reads *Man* and *man* differently, we always convert text to lower case, by default. For instance, if you used the text function in Excel that converts words to numbers, "man" is equal to 109, while "Man" is equal to 77. This is how Excel understands words.

Step 4: Punctuation removal.

Punctuation generally carries no meaning for machine learning algorithms. For instance, if we are studying *sentiment*, rather than assuming the meaning of a question mark or exclamation point in a sentence, we compare words for negative or positive sentiment using lexicons (more about this later).

Step 5: Remove non-alphabetic tokens.

These include *special characters*, and sometimes *numbers*. For some analyses, numbers are not important to the problem and for others they are. For instance, the frequency of the word, "2" or "two" may not be important is we are studying "gender bias." For instance, the number of times gender bias occurs may be obtained by tallying the number or frequency of bias cases or bias sentiment. However, if we are studying the number of products purchased by each gender, then numbers may be important. For instance, the number of men bought product A is 295, while the number of females who bought the product is 190. Without getting too deep into n-grams, the gender, product type, and quantity might be represented as *trigrams*.

Step 6: Remove the stop words.

Stop word or *stopword* removal is the process of filtering out useless data, which in natural language processing means useless words. A "useless word" is relative to the context, but generally includes common words like the articles "a," "an," "the," "that," or conjunctions like "but," "and," or other common word occurrences like pronouns. Pronouns are commonly removed because in NLP, the machine learning algorithm does not know which proper noun is referred to by "his," "she," "it," or "they."

Step 7: Perform stemming

Our next step is stemming the words or reducing words to a common *word stem*.

Definition 2.3. *Stemming is the process of reducing inflected or derived words to their word stem, base, or root form.*

These are generally a written word form, but the stem need not be identical to the morphological root of the word. It is usually sufficient that related words map to the same stem, even if this stem is not in itself a valid root. For instance, some algorithms reduce *argue*, *argued*, *argues*, *arguing*, and *argus* to the stem *argu*.

Step 8: Perform lemmatization

Lemmatizing the text is the process of grouping together the inflected forms of a word so they can be analyzed as a single item, identified by the word's lemma or dictionary form.

> ***Definition 2.4.*** *In computational linguistics, lemmatization is the algorithmic process of determining the lemma of a word based on its intended meaning.*

Unlike stemming, lemmatization depends on correctly identifying the intended part of speech and meaning of a word in a sentence, as well as within the larger context surrounding that sentence, such as neighboring sentences or even an entire document. For example, the word "better" has "good" as its lemma. This link is missed by stemming, as it requires a dictionary look-up. However, the word "walk" is the base form for word "walking," and this is matched in both stemming and lemmatization.

2.4 Load the Data

There are a number of ways to load data from different locations (i.e., URL, local drive, server, etc.) and of different formats (i.e., txt, csv, etc.). We can also load disjoint documents in one directory.

2.4.1 Loading Data in Python

Python has several options for loading or importing data, although *Pandas* is the most popular library.

Load CSV with Pandas Library

In Python or Jupyter Notebook, we can use functions from the *Pandas* library to load/import data. Sponsored by NumFOCUS, *Pandas* is an open source, Berkeley Software Distributions (BSD) licensed library providing high-performance, easy-to-use data structures and data analysis tools for the Python programming language. Besides the functionality to read and load a variety of file types, *Pandas* has functions to manipulate dataframes, missing data, reshaping data, slicing data, etc. For a full list of functionalities, see *https://pandas.pydata.org/*. For example, if we want to load data from an URL, we could designate the path of the URL and load the data that is in csv format, using:

```
# Load CSV from URL
import pandas as pd
data_source_url =
"https://raw.githubusercontent.com/kolaveridi/kaggle-Twitter-US-Airline-Sentiment-/master/Tweets.csv"
airline_tweets = pd.read_csv(data_source_url)
```

In Python, we can call on function from the *os* library the code might look like this:

```
# Load CSV file
from os import listdir
from os.path import isfile, join

mypath = "D:/Documents/Data/regs"
onlyfiles = [f for f in listdir(mypath) if isfile(join(mypath, f
))]
```

Load CSV with Python Standard Library

The Python API provides the *CSV* library and the function **reader()** that can be used to load CSV files. After loading ou data as CSV, we can use *Numpy* to put the data into an array. In order to use the data with machine learning algorithms, the dat must be in an array. The code might look like this:

```
# Load CSV (using python)
import csv
import numpy
filename = 'D:/Documents/Data/iris3.data.csv'
raw_data = open(filename, 'rt')
reader = csv.reader(raw_data, delimiter=',', quoting=csv.QUOTE_N
ONE)
x = list(reader)
data = numpy.array(x).astype('float')
print(data.shape)
```

Pandas is built on the Numpy package (discussed below). An advantage of using *pandas* is the function **pd.read_csv** returns a *pandas* **DataFrame** that you can immediately start summarizing and plotting. DataFrames allow you to store and manipulate tabular data in rows of observations and columns of variables.

If the data we are loading does not have header, we can modify the code so that header can be inserted and customized with an array of names. Then we can check the addition using **data.shape**.

```
# Load CSV using Pandas
import pandas
filename = 'D:/Documents/Data/iris3.data.csv'
names = ['Sepal.Length','Sepal.Width','Petal.Length','Petal.Widt
h','Species']
data = pandas.read_csv(filename, names=names)
print(data.shape)
```

Load CSV File with *NumPy*

Numpy provides a high-performance multidimensional array and basic tools to compute with and manipulate these arrays. We can also load our CSV data using *NumPy* and the **numpy.loadtxt()** function. Using this function assumes no header row and all data has the same format. The example below assumes that the file **iris3.csv** is not in your current working directory.

```
# Load CSV file
import numpy
filename = 'D:/Documents/Data/iris3.data.csv'
raw_data = open(filename, 'rt')
data = numpy.loadtxt(raw_data, delimiter = ",")
print(data.shape)
```

We can also use *Numpy* to load data from an url as follows:

```
# Load CSV from URL using NumPy
from numpy import loadtxt
from urllib.request import urlopen
url = 'https://raw.githubusercontent.com/jbrownlee/Datasets/mast
er/pima-indians-diabetes.data.csv'
raw_data = urlopen(url)
dataset = loadtxt(raw_data, delimiter = ",")
print(dataset.shape)
```

2.4.2 Lodaing Data in R

In R or R Studio, most of the data utilities are in libraries that load when we start.

read.csv()

If we have a comma-delimited file, we use R's `read.csv()` function. Using this function, we can read csv files from a local directory or url either by designating the path in the statement, for example:

```
path <- file.path("https://raw.githubusercontent.com/kolaveridi/
kaggle-Twitter-US-Airline-Sentiment-/master/Tweets.csv")
doc <- read.csv(path)
```

Or seting the working directory to where we store our data. For example, we can load the file document.csv stored at "D:/Documents/Data" with:

```
setwd("D:/Documents/Data")
doc <- read.csv(document.csv)
```

In R, the library that contains `read.csv()` is a utility, *utils*, and is loaded when R starts up.

Now, if we want to load a directory with multiple documents, we would code in R as:

```
cname <- file.path("D:/Documents/Data", "bible")
cname
dir(cname)
docs <- Corpus(DirSource(cname))
```

Also, in R, there are libraries, like *rpart*, that have "built-in" data sets. We load these like this:

```
library(rpart) # Library containing the dataset we want
data(kyphosis) # This statemeni loads the dataset
df <- kyphosis # Assigns the df as the dataset name
```

read.table()

If you have a .txt or a tab-delimited text file, you can easily import it with the basic R function `read.table()` from *utils*. For example:

```
df <- read.table("https://s3.amazonaws.com/assets.datacamp.com/b
log_assets/test.txt",  header = TRUE)
```

read.delim()

In case you have a file with a separator character that is different from a tab, a comma or a semicolon, you can always use the `read.delim()`

and `read.delim2()` functions. These are variants of the `read.table()` function, just like the `read.csv()` function.

readWorksheetFromFile()

We can also import Excel files using thei function from the CLConnect library. However, I find it simpler to **convert the Excel file to CSV format** and then use `read.csv()`. If we wish to use the Excel converter, from *XLConnect*, then a statement might appear as (Willems, 2018):

```
library(XLConnect)
df <- readWorksheetFromFile("<file name and extension>",
sheet = 1)
```

readHTMLTable()

Loading data from HTML tables into R is pretty straightforward using `readHTNLTable()`:

```
url <- "<a URL>" # Assign your URL to `url`
data_df <- readHTMLTable(url, which=3) # Read the HTML table
```

Read JSON Files Into R

To get JSON files into R, you first need to install or load the *rjson64* package. Later, we will show you how to install and load packages more efficiently, but for now we use (Willems, 2018):

```
install.packages("json64")
library(rjson) # Activate `rjson`
JsonData <- fromJSON(file= "<filename.json>" ) # Import data fro
m json file
```

The foreign-package

Using the *foreign* package, we can load files formatted as SAS, SPSS, Stata, Systsat, Minitab and more.

2.4.3 Loading Our Data

On your computer, create a folder "Data_Analytics" on your C or local drive to download the "text.zip" from my github directroy:

https://github.com/stricje1/Data

Then use the following code chunk to load your data into R Studio (the path you should use will be one on your computer's hard drive):

```
cname <- file.path("C:/Users/jeff/Documents/VIT_Course_Material/Data_Analytics_2018/data", "text")
cname <- file.path("C:/Users/username/Documents/Data_Analytics/data", "text")
dir(cname)
  [1] "A one-eyed man in the kingdom of the blind.txt"
  [2] "All Things Data.txt"
  [3] "Analytics and Statistics.txt"
  [4] "Analytics is it more than a buzzword.txt"
  [5] "Bayesian networks.txt"
  [6] "Big Data Analytics and Human Resources.txt"
  [7] "Big Data The Good the Bad and the Ugly.txt"
  [8] "Call Center Analytics.txt"
  [9] "Classification Trees using R.txt"
 [10] "Clouds clouds and more clouds.txt"
 [11] "Cluster Models.txt"
 [12] "Cyber-Threat Risk Assessment using R.txt"
 [13] "Data Scientist are Dead Long Live Data Science.txt"
 [14] "Do you like my Ensemble.txt"
 [15] "Free SAS.txt"
 [16] "Getting the Question Right.txt"
 [17] "What are Association Rules in Analytics.txt"
 [18] "Where_did_all_the_Teaching_Go.txt"
 [19] "Where_did_all_the_Thinking_Go.txt"
 [20] "Why_Stand_Many_Have_Fallen.txt"
```

We can check the document metadata via `DirSource()`,

```
docs <- Corpus(DirSource(cname))
```

```
Metadata:  corpus specific: 1, document level (indexed): 0
Content:  documents: 1
```

Now we examine the data we loaded using the `summry()` fucntion

```
summary(docs)
```

```
     Length  Class              Mode
[1]    2     PlainTextDocument  list
[2]    2     PlainTextDocument  list
[3]    2     PlainTextDocument  list
[4]    2     PlainTextDocument  list
[5]    2     PlainTextDocument  list
[6]    2     PlainTextDocument  list
[7]    2     PlainTextDocument  list
[8]    2     PlainTextDocument  list
[9]    2     PlainTextDocument  list
```

```
[10]  2      PlainTextDocument       list
[11]  2      PlainTextDocument       list
[12]  2      PlainTextDocument       list
[13]  2      PlainTextDocument       list
[14]  2      PlainTextDocument       list
[15]  2      PlainTextDocument       list
[16]  2      PlainTextDocument       list
[17]  2      PlainTextDocument       list
[18]  2      PlainTextDocument       list
[19]  2      PlainTextDocument       list
[20]  2      PlainTextDocument       list
```

Recall the data we have already inspected using the code snipit, `inspect(docs)`.

2.5 Convert to Lowercase

Converting all text to lowercase give every token (i.e. sentence or word) the same numeric value. Otherwise, the letters "H" and "h" have different values, so the word at the beginning , "Hold," of a sentence and the word, "hold," elsewhere in the sentencehave different values. Table 2-1, shows the numeric values for lower and uppercase forms of "hold."

Table 2-1. Lowercase and uppercase numeric values for "hold."

Char	Numval
h	104
o	111
l	108
d	100
(SUM)	**423**

Char	Numval
H	72
O	79
L	76
D	68
(SUM)	**295**

2.5.1 Convert to Lowercase in Python

In Python, `lower()` is a built-in method used for string manipulation. The `lower()` methods returns the lowercased string from the given string. It converts all uppercase characters to lowercase. If no uppercase characters exist, it returns the original string. For example, apllying `lower()` to the following string,

```
string = 'Big Data Analytics and Human Resources'
print(string.lower())
```

yields, the following output:

```
big data analytics and human resources
```

Of course, using upper() coverts the text to all caps, e.g.:

BIG DATA ANALYTICS AND HUMAN RESOURCES

2.5.2 Convert to Lowercase in R

StringR-package

Stingr from *tidyverse* package is popular choice, as all string functions begin with str() and are easy to remember. To convert to lower case, the code might look like this:

```
string <- 'Big Data Analytics and Human Resources'
str_to_lower(string, locale = "en")
```

When str_to_lower() is us applied to the string, it yields:

```
[1] "big data analytics and human resources"
```

Another way to convert the text to lowercase uses the *tm*-package. We can use arbitrary character processing functions as transformations as long as the function returns a text document. In this case we use content_transformer(), which provides a convenience wrapper to access andset the content of a document. Consequently, most text manipulation functions from base R can directly be used with this wrapper. This works for tolower() and other transformations like gsub(), which comes quite handy for a broad range of text manipulation tasks (Feinerer, 2019). Below, we use the *tm*-package's tolower() function on the dataset "crude" (data set holding 20 news articles)

```
library(tm)
data("crude")
tm_map(crude, tolower)
```

Next, we convert the corpus text (docs) to lowercase using tm_map and inspect our work so that we can see the lowercase conversion.

```
docs <- tm_map(docs, tolower)
inspect(docs[1])
```

```
A one-eyed man in the kingdom of the blind.txt
[a one-eyed man in the kingdom of the blind:\predicting the unpr
edictable\nâ\200œalmost nobodyâ\200\231s competent, paul. itâ\20
0\231s enough to make you cry to see how bad most people are at
their jobs. if you can do a half-assed job of anything, youâ\200
\231re a one-eyed man in the kingdom of the blind.â\200\235 â\20
0"kurt vonnegut, player piano\nabstract\nthis article is about p
redictive modeling. it explores the appropriateness of modeling
in general and predictive modeling in particular, as well as exa
mining some pitfalls. modeling is the process of formulating and
abstracting a representation of a real problem, based on simplif
ying assumptions. thus, no model is an exact representation of r
eality. said a different way, a model cannot fully represent a c
omplex problem, but can provide some insight into the problem an
d assist decision makers with applying solutions.]
```

2.6 Remove Numbers, Punctuation, & Whitespaces

Remove numbers if they are not relevant to your analyses. Usually, regular expressions are used to remove numbers.

2.6.1 *Remove Numbers in R*

We can remove numbers from text using the *quanteda* package, calling on the `tokens()` function from the *quanteda* package that takes one paramenter, in this case. Note that we can actually perform the removal of numbers and punctuation (and hyphens and symbols) in once code chunk. We will first perform the tasks one at a time and then put the together.

```
# the numbers are removed - no need to predict numbers
master_Tokens <- tokens(remove_numbers = TRUE)
```

2.6.2 *Remove Punctuation in R*

We remove punction marks, like colons and commas, is a similar manner, uning the `master_Tokenstokens()` function from *quanteda*.

```
# All punctuation marks are removed - no need to predict numbers
master_Tokenstokens(remove_punct = TRUE)
```

2.6.3 *Remove Whitespaces in R*

To remove whitespaces, we use R's basic trim function, `trimws().`

```
x<-doc$text
```

```
trimws(x)
trimws(x, "l") # Trim from left
trimws(x, "r") # Trim from right
```

We can do all of these tasks at once using this code chunck:

```
library(quanteda)
master_Tokens <- tokens(
  x = tolower(doc$text),
  remove_punct = TRUE,
  remove_numbers = TRUE, # Only if not needed for the analysis
  remove_hyphens = TRUE,
  remove_symbols = TRUE,
  remove_url = TRUE # if the text is from social media
)
```

2.6.4 Remove Numbers in Python

Removing numbers in Python is a simple process using the *re* library. To remove punctuation and whitespaces, we will rely on the functionality in the *string* library. The order in which tasks are performed seems to matter, intuitively. When we tokenize the text, we add quotes and commas, for instance, which we need. So, removing numbers, punctuation, and whitespaces should preceed tokenization.

```
import re
input_str = "Box A contains 3 red and 5 white balls, while Box B contains 4 red and 2 blue balls."
result = re.sub(r'\d+', '', input_str)
print(result)
```

To remove remove punctuation in Python, we use the

```
import string
input_str = "This &is [an] example? {of} string. with.? punctuation!!!! " # Sample string
result = input_str.translate(string.maketrans("",""),
         string.punctuation)
print(result)
```

To remove whitespaces in Python, we use the **input_str.strip()** function.

```
input_str = " \t a string example\t "
input_str = input_str.strip()
input_str
```

2.7 Tokenize the Text

As we saw in Section 2.3, tokenization is is the process of splitting the text into tokens like words, sentences, or paragraphs.

2.7.1 Tokenization in Python

The Natural Language Toolkit (NLTK) is a Python package for natural language processing. The *nltk* package offers a method to tokenize the text in Python, for example:

```
import pandas as pd
import nltk
df['text'] = df['text'].apply(word_tokenize)
```

We will discuss other ways to tokenize text when we visit the topic of vectorization in Chapter 3.

2.7.2 Tokeniztion in R

The *tokenizers* library for R provides the function `tokenize_words()`. The code mithg look loke this:

```
# Tokenization
if(!require(tokenizers)) install.packages("tokenizers")
Tweet_tk <- tokenize_words(tweets$tweets)
Tweet_tk
```

Another way to tokenize the text is using `unnest_tokens` from the *tidytext* package, using a *chain operator*, %>%. We could use the following code to tokenize the text:

```
# Tokenization
tidy_tweets<-tweets %>% unnest_tokens(word,tweets) %>% ungroup()
head(tidy_tweets,5)
```

We made use of the chain operator in the previous example, but we can use additional chain operators to perform several text transformations is one step.

```
tidy_tweets<-tweets %>% group_by(name,tweetid) %>% mutate(ln=row
_number()) %>% unnest_tokens(word,tweets) %>% ungroup()
head(tidy_tweets,5)
```

2.8 Word and Number Removal

Stop words are words that commonly appear in text, but do not typically carry significance for the meaning. For instance, "the," "for," and "it" are all considered stop words.

2.8.1 Remove Stopwords in R

There are two primary packages that I use with R: *Qyuanteda* and *tm*. With respect to efficiency, the quanteda-package is a very good choice. To remove stop words, we can use the function `tokens_select` with the arguments `stopwords('english')` and `selection = 'remove'`:

```
doc.tokens <- tokens_select(doc.tokens, stopwords('english'),
selection='remove')
```

tm-package

Returning to the *tm*-package we introduce in Section 2.5.2, it offers functionality for removing stopwords. The following code uses tm_map to remove stop words or stopwords:

```
crude <- tm_map(crude, removeWords, stopwords("english"))
```

It is often efficient to use the chain operator, %<%, to link several R code chunks, while we take care of stopwords. For instance, when we remove stopwords, the mutate function used in a chain preceding tokenization adds new variables and preserves existing ones, might look like this:

```
tidy_tweets<-tweets %>% group_by(name,tweetid) %>% mutate(ln=row
_number()) %>% unnest_tokens(word,tweets) %>% ungroup()
head(tidy_tweets,5)
```

2.8.2 Removing Stopwords in Python

Here, we use the *nltk* library to remove stopwords:

```
from nltk.corpus import stopwords
eng_stops = set(stopwords.words('english'))

# Remove stopwords from our documents: our_docs
stop = eng_stops
df['text'] = df['text'].apply(lambda x: [item for item in x if i
tem not in stop])
df['text']
```

2.8.3 Removing Stopwords from Our Text

Next, remove unnecessary words from the text using `tm_map()`:

```
docs <- tm_map(docs, removeNumbers)
docs <- tm_map(docs, removeWords, stopwords("English"))
docs <- tm_map(docs, removeWords, c("can", "should", "would",
"figure", "using", "will", "use", "now", "see", "may", "given",
"since", "want", "next", "like", "new", "one", "might", "without
"))
```

Now, combine words that should stay together using this user defined function:

```
for (j in seq(docs))
{
docs[[j]] <- gsub("data analytics", "data_analytics", docs[[j]])
docs[[j]] <- gsub("predictive models", "predictive_models",
docs[[j]])
docs[[j]] <- gsub("predictive analytics","predictive_analytics",
docs[[j]])
docs[[j]] <- gsub("data science", "data_science", docs[[j]])
docs[[j]] <- gsub("operations research", "operations_research",
docs[[j]])
docs[[j]] <- gsub("chi-square", "chi_square", docs[[j]])
}
```

2.9 Stemming the Text

I have heard stemming described as a crude heuristic process that chops off the ends of words in hope of getting all derivative of a word to the base word. For example, the words *catching*, *catches*, and *caught* would be reduced to *catch*. And though this might be "crude," it is a method that often works, without the added expense of lemmatization.

Now, you might be asking yourself, what we mean by expensive. In the contest of NLP, expensive refers to the time, resourses, and cost of solving a problem or implementing a solution. This takes in to consideration, man-hours, computer time, data access, and so on. Stemming is relatively cheap and fast. However, stemming is optional, and if you are lemmatizing your text, stemming can be detrimental to your analysis.

2.9.1 Stemming in R

The main two algorithms are *Porter stemming algorithm* (removes common morphological and inflexional endings from words) and *Lancaster stemming algorithm* (a more aggressive stemming algorithm). Here we will use the function **tokens_wordstem()**, which is another function from the *quanteda* package.

```
stemed_words <- tokens_wordstem(master_Tokens, language = "engli
sh")
```

Moreover, stemming is not looking for a correct root word, it's just loppimh off endning, as see with the word "crazi:"

```
[8] "get"          "hung"          "up"            "on"            "crazi"
```

2.9.2 Stemming in Python

Porter stemming is a Python library, *PorterStemmer*, that implements the Porter stemming algorithm, and it is a nltk companion.

```
from nltk.stem import PorterStemmer
from nltk.tokenize import word_tokenize
stemmer= PorterStemmer()
input_str= "There are several types of stemming algorithms."
input_str=word_tokenize(input_str)
for word in input_str:
    print(stemmer.stem(word))
```

2.9.3 Stemming our text

In linguistic morphology and information retrieval, stemming is the process of reducing inflected (or sometimes derived) words to their word stem, base or root form. For example, as *car*, *car*, car's, and *cars'* might have a root form *car*. The stem need not be identical to the morphological root of the word; it is usually sufficient that related words map to the same stem, even if this stem is not in itself a valid root.

2.10 Text Lemmatization

The aim of lemmatization, like stemming, is to reduce inflectional forms to a common base form. As opposed to stemming, lemmatization does not simply chop off inflections. Instead it uses lexical knowledge bases to get the correct base forms of words.

2.10.1 Lemmatization in Python

Lemmatization is supported by these libraries, NLTK (WordNet Lemmatizer), spaCy, TextBlob, Pattern, gensim, Stanford CoreNLP, Memory-Based Shallow Parser (MBSP), Apache OpenNLP, Apache Lucene, General Architecture for Text Engineering (GATE), Illinois Lemmatizer, and DKPro Core.

```
import pandas as pd
from nltk.stem import WordNetLemmatizer
from nltk.tokenize import word_tokenize
corpus = pd.read_csv("D:/Documents/Data/case_study_data_tiny.csv
")
# using list comprehension
text = ' '.join(map(str, corpus['text']))
lemmatizer= WordNetLemmatizer()
input_str = text
input_str = word_tokenize(input_str)
for word in input_str:
    print(lemmatizer.lemmatize(word))
```

2.10.2 Lemmatization in R

Lemmatization takes into consideration the morphological analysis of the words. To do so, it is necessary to have detailed dictionaries which the algorithm can look through to link the form back to its lemma. The key to this methodology is linguistics. To extract the proper lemma, it is necessary to look at the morphological analysis of each word.

Use `lemmatize_words()` from the *textstem* package for lemmatizing a vector of individual words. Use `lemmatize_strings()` to lemmatize words within a string without extracting the words.

```
# Load or install required packages
if(!require(textstem)) install.packages("textstem")
# Define check_dictionary in lemmatize_words function (it's miss
ing in the source code)
check_dictionary <- function(x){
  if(anyDuplicated(x[[1]]) > 0) stop("Duplicate tokens found in
column one of the lemma dictionary.")
}
# Define numreg in lemmatize_strings function (it's missing in t
he source code)
numreg <-  "(?<=^| )[-.]*\\d+(?:\\.\\d+)?(?= |\\.?$)|\\d+(?:,\\d
{3})+(\\.\\d+)*"
```

```
# Words to lemmatize
z <- c("the", NA, 'doggies', ',', 'well', 'they', "aren\'t", 'Jo
yfully', 'running', '.')
lemmatize_words(z)
```

```
[1] "the" NA "doggy" "," "good" "they" "aren't" "Joyfully" "run"
"."
```

```
# String to lemmatize
x <- c(
  'the dirtier dog has eaten the pies',
  'that shameful pooch is tricky and sneaky',
  "He opened and then reopened the food bag",
  'There are skies of blue and red roses too!',
  NA,
  "The doggies, well they aren't joyfully running.",
  "The daddies are coming over...",
  "This is 34.546 above"
)
## Default lexicon::hash_lemmas dictionary
lemmatize_strings(x)
```

```
[1] "the dirty dog have eat the pie"
[2] "that shameful pooch be tricky and sneaky"
[3] "He open and then reopen the food bag"
[4] "There be sky of blue and red rose too!"
[5] NA
[6] "The doggy, good they aren't joyfully run."
[7] "The daddy be come over..."
[8] "This be 34. 546 above"
```

The next snippet of code lemmatizes a longer string, the first chapter of 1 John in the Holy Bible (output not shown).

```
library(readtext)
library(textstem)
library(stringi)
corpus <- readtext::readtext("D:/Documents/Data/reviews.csv")
stringi::stri_locale_set("en_GB")
corpus$text <- stringi::stri_replace_all(corpus$text, "", regex
= "\\[[0-9]+\\]")
corpus$text <- stringi::stri_replace_all(corpus$text, "", fixed
= "[citation needed]")
doc.lemma <- lemmatize_words(corpus$text, dictionary = lexicon::
hash_lemmas)
#doc.lemma
lemmatize_strings(doc.lemma)
```

2.11 Document Matrix Creation

A term-document matrix is an important representation for text analytics. In a document-term matrix, rows correspond to documents in the collection and columns correspond to terms.

Definition 2.5. *A document-term matrix or term-document matrix is a mathematical matrix that describes the frequency of terms that occur in a collection of documents.*

Logically, each row of the matrix is a document vector, with one column for every term in the entire corpus. A sparse matrix may not contain a given term. The value in each cell of the matrix is the term frequency.

2.11.1 Create Document Matrices

In these setps we will prepare the documents for analysis. First, we will put the text into a term-document matrix and view it:

```
tdm <- TermDocumentMatrix(docs)
tdm
```

```
<<TermDocumentMatrix (terms: 3971, documents: 20)>>
 Non-/sparse entries: 7178/72242
 Sparsity           : 91%
 Maximal term length: 18
 Weighting          : term frequency (tf)
```

Second, create document-term matrix and view it:

```
dtm <- DocumentTermMatrix(docs)
dtm
```

```
<<DocumentTermMatrix (documents: 20, terms: 3971)>>
 Non-/sparse entries: 7178/72242
 Sparsity           : 91%
 Maximal term length: 18
 Weighting          : term frequency (tf)
```

Next, organize the terms by their frequency:

```
freq <- colSums(as.matrix(dtm))
length(freq)
ord <- order(freq)
```

Now, put it into a matrix and save it to your working directory:

```
m <- as.matrix(dtm)
dim(m)
write.csv(m, file="dtm.csv")
```

```
 [1]   20 3971
```

Remove sparse terms. This makes a matrix that is a maximum of 10% empty space.

```
dtms <- removeSparseTerms(dtm, 0.1)
inspect(dtms)
```

```
<<DocumentTermMatrix (documents: 20, terms: 0)>>
 Non-/sparse entries: 0/0
 Sparsity           : 100%
 Maximal term length: 0
 Weighting          : term frequency (tf)
 Sample             :
                                                          Terms
 Docs
   A one-eyed man in the kingdom of the blind.txt
   All Things Data.txt
   Analytics and Statistics.txt
   Analytics is it more than a buzzword.txt
   Bayesian networks.txt
   Big Data Analytics and Human Resources.txt
   Big Data The Good the Bad and the Ugly.txt
   Call Center Analytics.txt
   Classification Trees using R.txt
   Clouds clouds and more clouds.txt
   Cluster Models.txt
   Cyber-Threat Risk Assessment using R.txt
   Data Scientist are Dead Long Live Data Science.txt
   Do you like my Ensemble.txt
   Free SAS.txt
   Getting the Question Right.txt
   What are Association Rules in Analytics.txt
   Where_did_all_the_Teaching_Go.txt
   Where_did_all_the_Thinking_Go.txt
   Why_Stand_Many_Have_Fallen.txt
```

2.11.2 Develop Frequency Distributions

Next, we check some of the frequency counts using the *knitr* package and `kable()` function.. There are a lot of terms, so for now, we just check out some of the most and least frequently occurring words, as well as check out the frequency of frequencies.

```
knitr::kable(findFreqTerms(dtm, lowfreq=60), caption = 'Frequent
Words')
```

Frequent Words

x
analytics
data
model
true
classification
tree
dendrogram

```
knitr::kable(freq[1:10], caption = 'Frequency Counts')
```

Frequency Counts

	x
data	198
analytics	104
true	68
model	67
classification	66
dendrogram	64
tree	62
branches	59
clustering	59
models	57

```
findFreqTerms(dtm, lowfreq >= 100)
```

```
[1] "data"      "analytics"
```

2.11.3 Visualizing the Results

Now, we plot frequently occurrung words that appear at least 50 times. This is depicted as histogram in Figure 2-1.

```
wf <- data.frame(word=names(freq), freq = freq)
datatable(wf, options = list(pageLength = 5, scroll = '400px'))
p <- ggplot(subset(wf, freq > 30), aes(word, freq))
p <- p + geom_bar(stat="identity")
p <- p + theme(axis.text.x=element_text(angle = 45, hjust = 1))
p
```

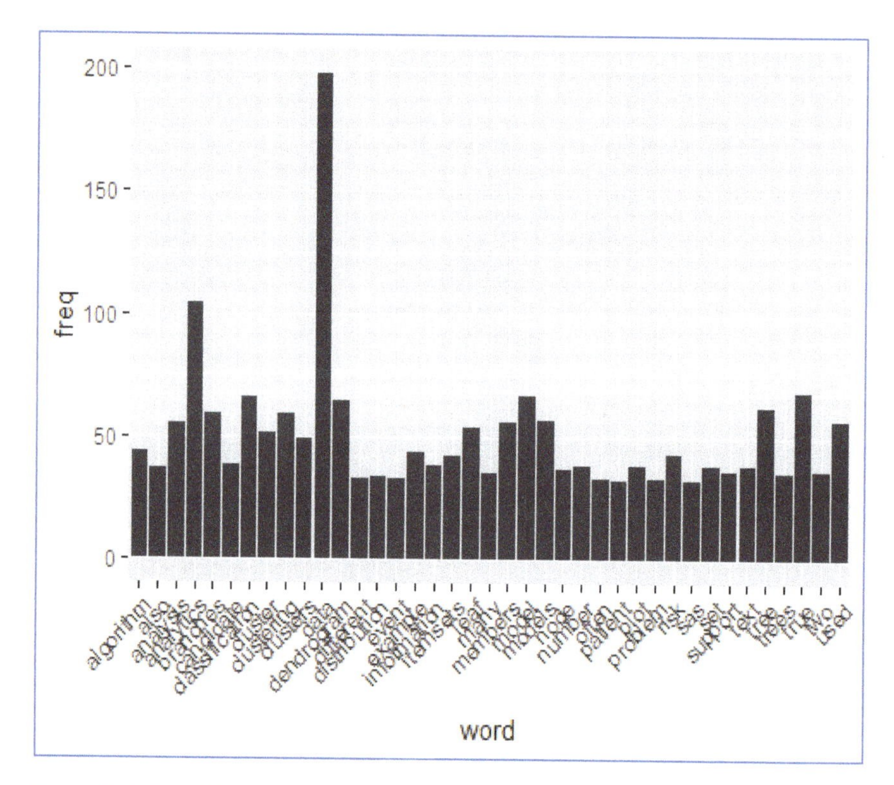

Figure 2-1. Word frequency distribution (histogram).

2.11.4 Find correlations

Now, we find correlations in the text.

```
findAssocs(dtm, c("question" , "analysis"), corlimit = 0.98) # s
pecifying a correlation limit of 0.98
```

```
$`question`
   achieve   behaviors     brainer        cart
      0.99        0.99        0.99        0.99
  currency        dave      detect     deviceÃ
      0.99        0.99        0.99        0.99
  dialogue    director  downstream   expertise
      0.99        0.99        0.99        0.99
  failures     forever       horse  investment
      0.99        0.99        0.99        0.99
      john        keys     knowing     mention
      0.99        0.99        0.99        0.99
    months  phenomenon     realize      recipe
      0.99        0.99        0.99        0.99
reiterated         ret    robinson       rolls
      0.99        0.99        0.99        0.99
```

```
    roske      roskeÃ       seen      slides 
     0.99       0.99        0.99        0.99 
   staffs      stake       stems    temporal 
     0.99       0.99        0.99        0.99 
   timing      twice       vince      wallet 
     0.99       0.99        0.99        0.99 
$analysis
 numeric(0)
```

```
findAssocs(dtms, "contrast", corlimit = 0.90) # specifying a correlation limit of 0.95
```

2.11.5 Using Wordclouds to Visualize Results

A wordcloud is an image composed of words used in a particular text or subject, in which the size of each word indicates its frequency or importance. That is, the larger the word is, the more frequently it occurs in the text. The next plot is for words, using a wordcloud, that occur at least 50 times. This is representd by a *word cloud* or *wordcloud* (from wordcloud2) in Figure 2-2.

```
wordcloud2(subset(wf, freq > 50))
```

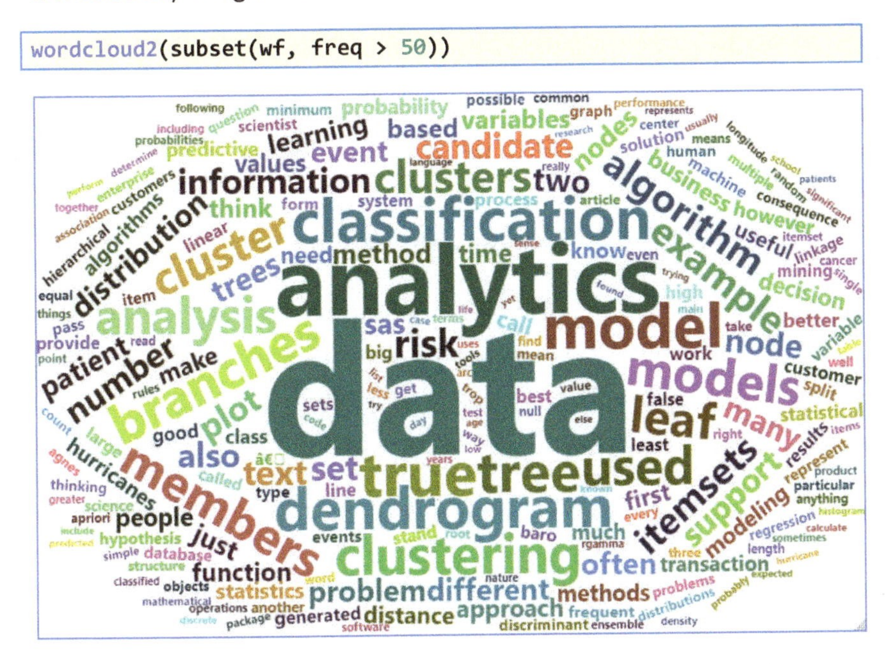

Figure 2-2. Word cloud representation of word frequencies, where "data" has the highest count.

2.12 Tidy Text Analytics I

2.12.1 *Tidy Format*

The tidy format is described by Hadley Wickham (Wickham 2014), with the following specific structure: - Each variable is a column - Each observation is a row - Each type of observational unit is a table

Definition 2.6. *The tidy text format is a table with one token per row.*

This one-token-per-row structure is in contrast to the ways text is often stored in current analyses, perhaps as strings or in a document-term matrix.

2.12.2 *Tidy Text Format*

Applying this to text, we can say that the tidy text format is table with one-token-per-row. A token is usually taken to be a word and is a meaningful unit of text that we are interested in using for analysis. The process of splitting text into tokens is called tokenization. In addition to words, the token that is stored in each row can be an n-gram, sentence, or paragraph. In the tidytext package, functionality is provided to tokenize by commonly used units of text like these and convert to a one-term-per-row format. However, not that the one-token-per-row structure of tidy text is in contrast to the usual methods of storing text for analyses, like strings or in a document-term matrix.

Tidy text data sets can be manipulated with a standard set of "tidy" tools, including popular packages such as tidyr (Wickham 2016), dplyr (Wickham and Francois 2016), broom (Robinson 2017), and ggplot2 (Wickham 2009).

2.12.3 *Contrasting tidy text with other data structures*

Given the unique appraoch of tidy text, it is worth contrasting with the ways text is often stored in text mining approaches. - **String**: Text can, of course, be stored as strings, i.e., character vectors, within R, and often text data is first read into memory in this form. - **Corpus**: These types of objects typically contain raw strings annotated with additional metadata and details. - **Document-term matrix**: This is a sparse matrix describing a collection (i.e., a corpus) of documents with one row for each document and one column for each term.

```
#Insatlling Required Libraries
if(!require(tidytext)) install.packages("tidytext")
if(!require(tidyverse)) install.packages("tidyverse")
```

```
Loading required package: tidyverse
-- Attaching packages -------------------------------------------
------------------------- tidyverse 1.2.1 --
v ggplot2 3.0.0     v purrr   0.2.5
v tibble  1.4.2     v dplyr   0.7.6
v tidyr   0.8.1     v stringr 1.3.1
v readr   1.1.1     v forcats 0.3.0
-- Conflicts ----------------------------------------------------
-------------------- tidyverse_conflicts() --
x dplyr::filter() masks stats::filter()
x dplyr::lag()    masks stats::lag()
```

```
if(!require(dplyr)) install.packages("dplyr")
if(!require(tidyr)) install.packages("tidyr")
if(!require(tibble)) install.packages("tibble")
if(!require(readr)) install.packages("readr")
if(!require(broom)) install.packages("broom")
if(!require(ggplot2)) install.packages("ggplot2")
```

2.12.4 The unnest_tokens function

We will use an example to look at the basic operations required for "tidying" text. Here we have the first stanz of the song "Hallelujah," released by Leonard Cohen (1984) and we place it into an array.

```
text_sample <- c("Now I've heard there was a secret chord",
    "That David played, and it pleased the Lord",
    "But you don't really care for music, do you?",
    "It goes like this",
    "The fourth, the fifth",
    "The minor fall, the major lift",
    "The baffled king composing Hallelujah"
    )
```

```
Text_sample
 [1] "Now I've heard there was a secret chord"
 [2] "That David played, and it pleased the Lord"
 [3] "But you don't really care for music, do you?"
 [4] "It goes like this"
 [5] "The fourth, the fifth"
 [6] "The minor fall, the major lift"
 [7] "The baffled king composing Hallelujah"
```

This is a typical character vector that we might want to analyze. In order to turn it into a tidy text dataset, we first need to put it into a data frame.

Nesting creates a list-column of data frames; **unnesting** flattens it back out into regular columns. Nesting is implicitly a summarising operation: we get one row for each group defined by the non-nested columns. This is useful in conjunction with other summaries that work with whole datasets, most notably models.

We will use the data-frame function from the dplyr package to accomplish this.

```
library(dplyr)
text_sample_df <- data_frame(line = 1:7, text = text_sample)
text_sample_df
```

```
# A tibble: 7 x 2
   line text
  <int> <chr>
1     1 Now I've heard there was a secret chord
2     2 That David played, and it pleased the Lord
3     3 But you don't really care for music, do you?
4     4 It goes like this
5     5 The fourth, the fifth
6     6 The minor fall, the major lift
7     7 The baffled king composing Hallelujah
```

2.12.5 Tibbles

Notice that the data frame printed out as a "tibble." Within R, a tibble is a modern class of data frame available in the *dplyr* and *tibble* packages with the following features: - has a convenient print method - will not convert strings to factors - does not use row names. Tibbles are good to use with tidy tools.

Definition 2.7. *Tibbles are a modern take on data frames. They keep the features that have stood the test of time, and drop the features that used to be convenient but are now frustrating (i.e. converting character vectors to factors).*

However, this text data frame is not yet compatible with tidy text analysis. Since each row is made up of multiple words, we cannot filter out words or count which occur most frequently. We need to convert this so that it has one-token-per-row. Later, when we work with multiple documents, we will need one-token-per-document-per-row.

One way to do this is to unnest the tibble. Here, we define the text of Pride and Prejudice as a tibble and unnest it by words.

```r
library(dplyr)
library(janeaustenr)
library(tidytext)
d <- tibble(txt = prideprejudice)
d %>% unnest_tokens(word, txt)
```

```
# A tibble: 122,204 x 1
   word
   <chr>
 1 pride
 2 and
 3 prejudice
 4 by
 5 jane
 6 austen
 7 chapter
 8 1
 9 it
10 is
# ... with 122,194 more rows
```

2.12.6 Tokenization

As we have stated, token is a meaningful unit of text that we are interested in using for further analysis. Tokenization is the process of splitting text into tokens. To obtain the tidy text framework, we need to break the text into individual tokens and transform the text to a tidy data structure. To do this, we use the *tidytext* function **unnest_tokens()**.

```r
library(tidytext)
text_sample_df %>% unnest_tokens(word, text)
```

```
# A tibble: 44 x 2
    line word
   <int> <chr>
 1     1 now
 2     1 i've
 3     1 heard
 4     1 there
 5     1 was
 6     1 a
 7     1 secret
 8     1 chord
 9     2 that
10     2 david
# ... with 34 more rows
```

The two basic arguments to `unnest_tokens` are the column names, "word" and "text." First, we have the output column name "word" that will be created as the text is unnested into it. Second,we have the input column "text" that the text comes from (Recall that text_sample_df above has a column called text that contains the data of interest).

After using `unnest_tokens()`, we split each row so that there is one token (word) in each row of the new data frame, which is the default tokenization in `unnest_tokens()`. Also observe that the line number each word came from are retained and punctuation has been stripped. Moreover, `unnest_tokens()` converts the tokens to lowercase by default, which makes them easier to combine or compare with other datasets.

2.13 Example - Indian Philosophy

After teaching data analytics at the Vellore Institute of Technology (VIT), at Vellore India, in 2016, my friend and Dean of the School for Information Technology Engineering (SITE) gave me a two-volume set of books on "Indian Philosophy." In order to enhance my understanding of what I had read, I performed text analytics on them, and repeat that here as an example.

First, we put ind texts to a data frame, a tibble.

```
ind_words <- tibble(file = paste0("~/tidy_text/",
    c("Indian_Philosophy_Part_I.txt", "Indian_Philosophy_Part_II
.txt"))) %>%
    mutate(text = map(file, read_lines))
ind_words
```

```
# A tibble: 2 x 2
  file                                        text
  <chr>                                       <list>
1 ~/tidy_text/Indian_Philosophy_Part_I.txt  <chr [13,715]>
2 ~/tidy_text/Indian_Philosophy_Part_II.txt <chr [19,984]>
```

The resulting tibble has a variable file that is the name of the file that created that row and a list-column of the text of that file.

We want to `unnest()` that tibble, remove the lines that are LaTeX crude and compute a line number.

```
ind_words <- ind_words %>%
  unnest() %>%
  filter(text != "%!TEX root = ind.tex") %>%
  filter(!str_detect(text, "^(\\\\[A-Z,a-z])"),
         text != "") %>%
  mutate(line_number = 1:n(),
         file = str_sub(basename(file), 1, -5))
```

Now we have a tibble with file giving us the chapter, text giving us the line of text from the text files and `line_number` giving a counter of the number of lines since the start of the ind texts.

Now we want to tokenize (strip each word of any formatting and reduce down to the root word, if possible). This is easy with `unnest_tokens()`. I played around with the results and came up with some other words that needed to be deleted (stats terms like *ci* or *p*, LaTeX terms like _*i* or tabular and references/numbers).

```
ind_words <- ind_words %>%
  unnest_tokens(word, text) %>%
  filter(!str_detect(word, "[0-9]"),
         word != "fismanreview",
         word != "multicolumn",
         word != "p",
         word != "_i",
         word != "c",
         word != "ci",
         word != "al",
         word != "dowellsars",
         word != "h",
         word != "tabular",
         word != "t",
         word != "ref",
         word != "cite",
         !str_detect(word, "[a-z]_"),
         !str_detect(word, ":"),
         word != "bar",
         word != "emph",
         !str_detect(word, "textless"))
ind_words
```

```
# A tibble: 250,164 x 3
   file                       line_number word
   <chr>                            <int> <chr>
 1 Indian_Philosophy_Part_I             1 indian
 2 Indian_Philosophy_Part_I             1 philosophy
 3 Indian_Philosophy_Part_I             1 part
```

```
 4 Indian_Philosophy_Part_I            1 i
 5 Indian_Philosophy_Part_I            2 chapter
 6 Indian_Philosophy_Part_I            2 i
 7 Indian_Philosophy_Part_I            3 introduction
 8 Indian_Philosophy_Part_I            4 general
 9 Indian_Philosophy_Part_I            4 characteristics
10 Indian_Philosophy_Part_I            4 of
# ... with 250,154 more rows
```

Now to compute the sentiment using the words written per line in the ind texts. *tidytext* comes with three **sentiment lexicons**, AFFIN, Bing and Loughran. **AFFIN** provides a score ranging from −5 (very negative) to +5 (very positive) for 2,476 words. Bing Liu maintains and freely distributes a sentiment lexicon (the **Bing** lexicon) consisting of lists of strings, including 2006 positive words and 4783 negative words (Bing & Minqing, 2004).

Loughran is used to determine which tokens (collections of characters) are classified as words. Also contains sentiment classifications, counts across all filings, and other useful information, such as eight sentiment category identifiers, Harvard Word List identifier, number of syllables, and source for each word. (Loughran & McDonald, 2016). The Master Dictionary also tabulates all of the sentiment word lists. None of these account for negation ("I'm not sad," is a negative sentiment, not a positive one).

Using the loughran lexicon, with the `get_sentiments()` function, let's see how the emotions of my words change over the two texts. Figure 2-3 shows ten frequent words share by the corpus and their frequencies of occurance.

```
ind_words %>%
  inner_join(get_sentiments("loughran")) %>%
  group_by(index = line_number %/% 25, file, sentiment) %>%
  summarize(n = n()) %>%
  ggplot(aes(x = index, y = n, fill = file)) +
  geom_bar(stat = "identity", alpha = 0.8) +
  facet_wrap(~ sentiment, ncol = 5)
```

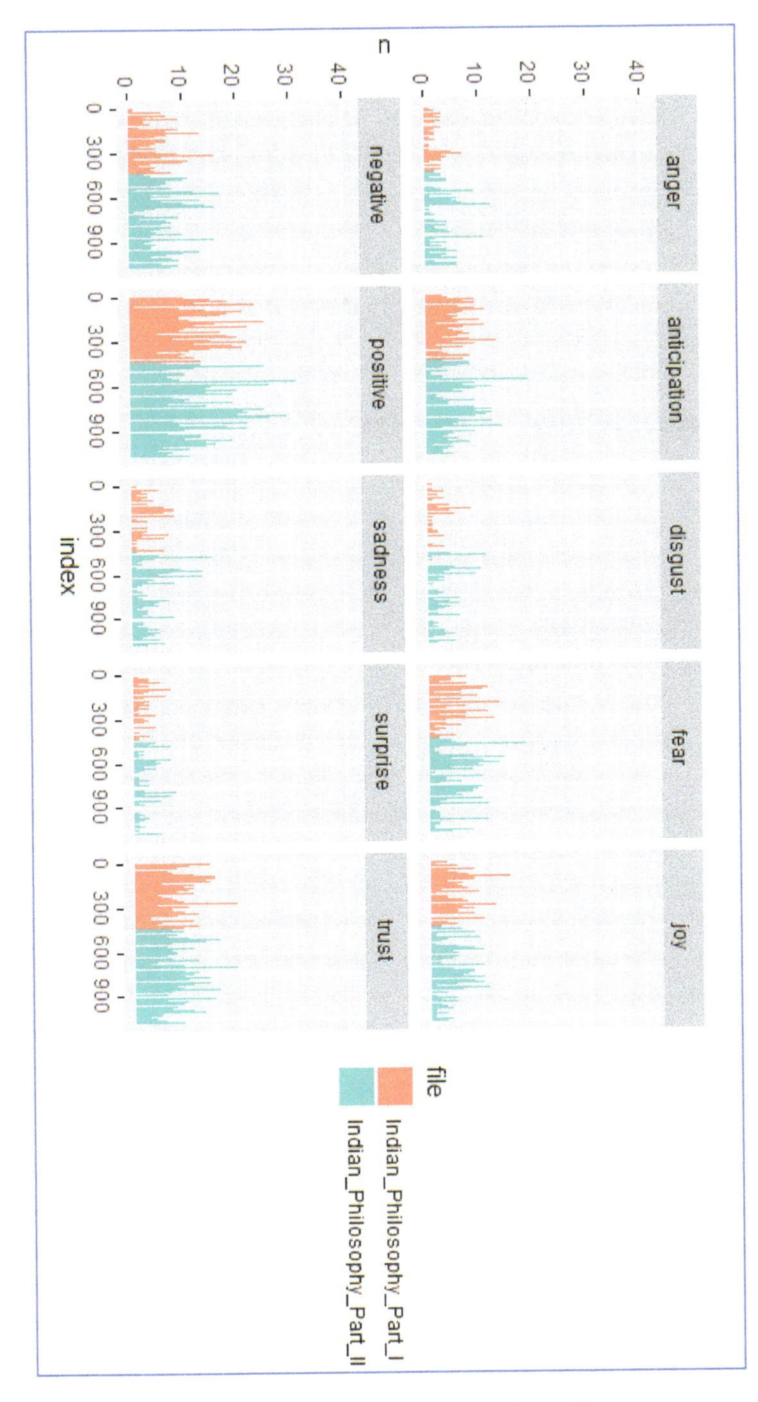

Figure 2-3. Histograms of ten English words from the Indian corpus.

Both texts are more positive than negative and represent trust fairly well. It looks like "disgust" and "sadness" are minimized. We can use the bing and afinn lexicons to look at how the sentiment of the words changed over the course of the thesis. This is depicted in Figure 2-4.

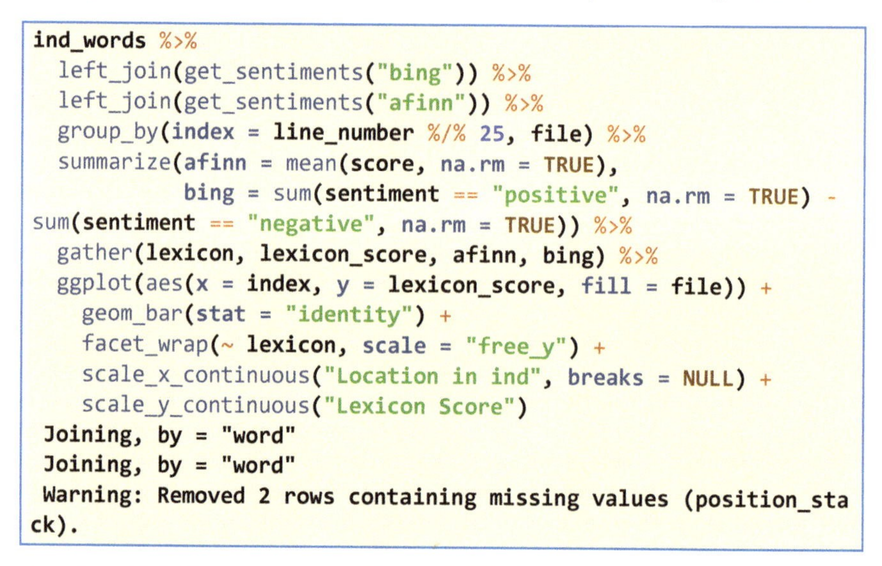

```
ind_words %>%
  left_join(get_sentiments("bing")) %>%
  left_join(get_sentiments("afinn")) %>%
  group_by(index = line_number %/% 25, file) %>%
  summarize(afinn = mean(score, na.rm = TRUE),
            bing = sum(sentiment == "positive", na.rm = TRUE) -
sum(sentiment == "negative", na.rm = TRUE)) %>%
  gather(lexicon, lexicon_score, afinn, bing) %>%
  ggplot(aes(x = index, y = lexicon_score, fill = file)) +
    geom_bar(stat = "identity") +
    facet_wrap(~ lexicon, scale = "free_y") +
    scale_x_continuous("Location in ind", breaks = NULL) +
    scale_y_continuous("Lexicon Score")
 Joining, by = "word"
 Joining, by = "word"
 Warning: Removed 2 rows containing missing values (position_sta
ck).
```

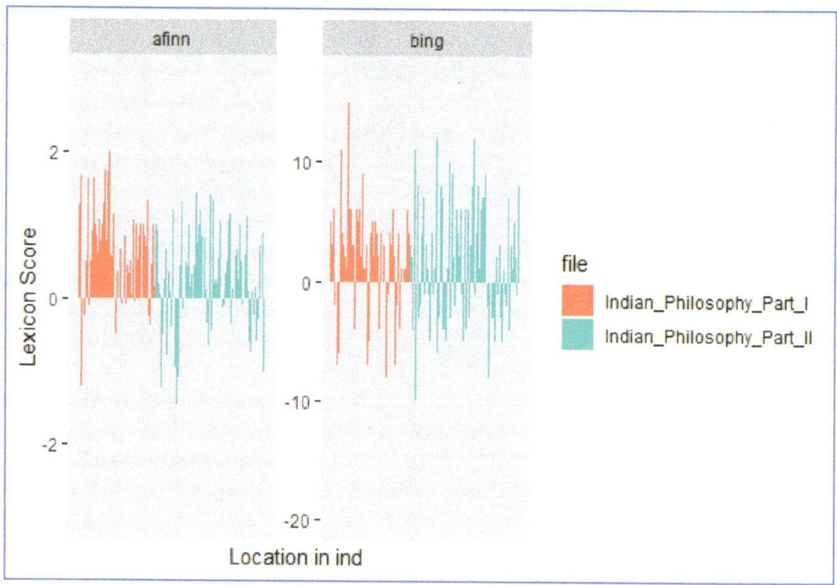

Figure 2-4. Comparison of lexicon scores for the two volume Indian corpus.

```
ind_words
```

```
# A tibble: 250,164 x 3
   file                         line_number word
   <chr>                              <int> <chr>
 1 Indian_Philosophy_Part_I               1 indian
 2 Indian_Philosophy_Part_I               1 philosophy
 3 Indian_Philosophy_Part_I               1 part
 4 Indian_Philosophy_Part_I               1 i
 5 Indian_Philosophy_Part_I               2 chapter
 6 Indian_Philosophy_Part_I               2 i
 7 Indian_Philosophy_Part_I               3 introduction
 8 Indian_Philosophy_Part_I               4 general
 9 Indian_Philosophy_Part_I               4 characteristics
10 Indian_Philosophy_Part_I               4 of
# ... with 250,154 more rows
```

Looking at the two lexicon's scoring of my books, the affin lexicon seems a little more stable if we assume local correlation of sentiments is likely. The scores show that all three text are much more positive than negative.

2.13.1 Filter for negative words

Next, we apply a filter for negative words in order to distingish between positive and negative in a wordcloud, for example.

```
bingnegative <- get_sentiments("bing") %>%
    filter(sentiment == "negative")
```

The, we get a word count.

```
wordcounts <- ind_words %>%
    group_by(index = line_number %/% 25, file) %>%
    summarize(words = n())
wordcounts
```

```
# A tibble: 1,081 x 3
# Groups:   index [?]
   index file                     words
   <dbl> <chr>                    <int>
 1     0 Indian_Philosophy_Part_I   187
 2     1 Indian_Philosophy_Part_I   240
 3     2 Indian_Philosophy_Part_I   251
 4     3 Indian_Philosophy_Part_I   227
 5     4 Indian_Philosophy_Part_I   215
 6     5 Indian_Philosophy_Part_I   239
 7     6 Indian_Philosophy_Part_I   241
 8     7 Indian_Philosophy_Part_I   239
```

```
 9     8 Indian_Philosophy_Part_I   228
10     9 Indian_Philosophy_Part_I   221
# ... with 1,071 more rows
```

2.13.2 Build a cloud chart

We use the previus code chunk results to build a wordcloud. This is shwn in Figure 2-5, using the *wordcloud* package.

```
library(wordcloud)
ind_words %>%
  anti_join(stop_words) %>%
  count(word) %>%
  with(wordcloud(word, n, max.words = 100))
```

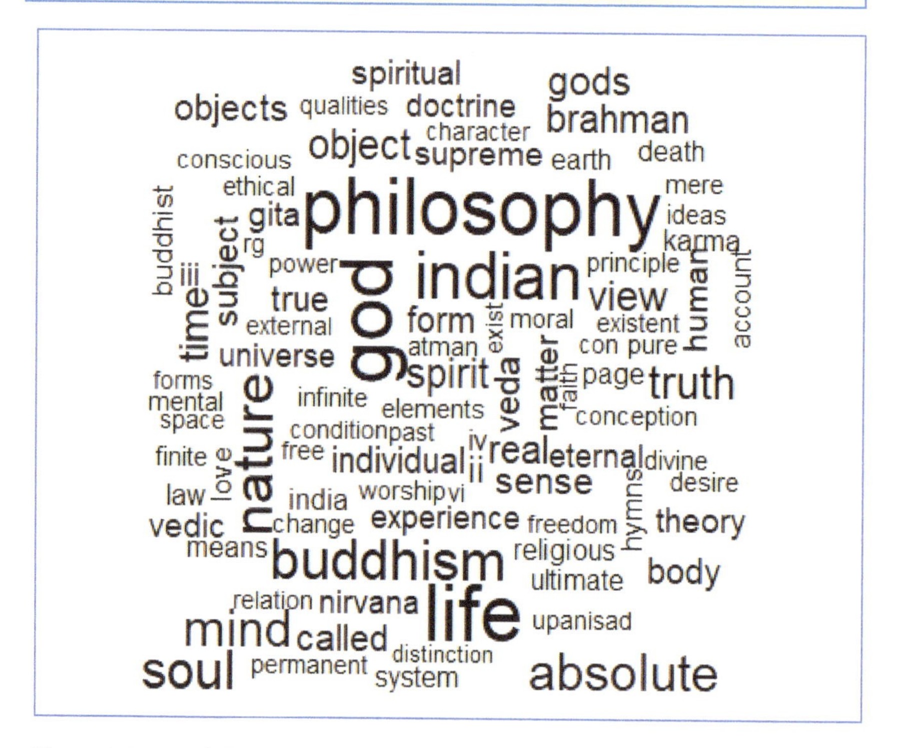

Figure 2-5. Word cloud representing word frequencies in the Indian corpus.

Finally, with the *reshape2* package, we build a contrasting cloud chart shown in Figure 2-6.

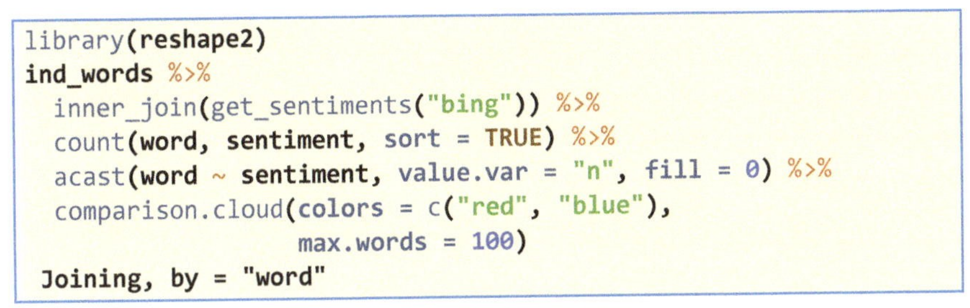

```
library(reshape2)
ind_words %>%
  inner_join(get_sentiments("bing")) %>%
  count(word, sentiment, sort = TRUE) %>%
  acast(word ~ sentiment, value.var = "n", fill = 0) %>%
  comparison.cloud(colors = c("red", "blue"),
                   max.words = 100)
 Joining, by = "word"
```

Figure 2-6. Contrasting cloud chart for the Indian corpus.

END OF SCRIPT

2.14 Tidy Text Analytics II

First, we load the necessary libraries, including some new ones like *sentimentr*, *lubridate*, *purrr*, *janeaustenr*, *igraph*, and *plotrix*.

```
if(!require(tm)) install.packages("tm")
if(!require(tidytext)) install.packages("tidytext")
if(!require(tidyverse))
install.packages("tidyverse")
if(!require(stringr)) install.packages("stringr")
if(!require(dplyr)) install.packages("dplyr")
if(!require(tidyr)) install.packages("tidyr")
if(!require(janeaustenr))
install.packages("janeaustenr")
if(!require(purrr)) install.packages("purrr")
if(!require(sentimentr))
install.packages("sentiment")
if(!require(readr)) install.packages("readr")
if(!require(wordcloud))
install.packages("wordcloud")
if(!require(lubridate))
install.packages("lubridate")
if(!require(ggplot2)) install.packages("ggplot2")
if(!require(ggraph)) install.packages("ggraph")
if(!require(igraph)) install.packages("igraph")
if(!require(plotrix)) install.packages("plotrix")
```

2.14.1 Tidy Text

As already covered, tidy text is a format useful for several typse of text mining problems, including sentiment and topic analysis. Some advantages of the tidy text format are: - keeps one token (typically a word) in each row - keeps each variable (such as a document or chapter) in a column. - when your data is tidy, you can use a common set of tools for exploring and visualizing them

This frees you from struggling to get your data into the right format for each task and lets you focus on the questions you want to ask.

Step 1: put the text into a data frame

- the `c()` function returns a vector (a one-dimensional array)
- `Paste0()` concatenates strings without spaces
- the `data_frame()` function is used for storing data tables

- The map() function transforms the text by applying a function to each element and returning a vector the same length
- read_lines() reads up to n_max lines from a file... read_lines_raw() produces a list of raw vectors
- the mutate() function adds new variables and preserves existing

```
quantum_words<-
 data_frame(file =
   paste0("C:\\Users\\jeff\\Documents\\VIT_Course_Material\\Data
_Analytics_2018\\data\\",
   c("quantum_phaith.txt",
     "quantum_hope.txt",
     "quantum_love.txt"))) %>%
   mutate(text = map(file, read_lines))
```

Step 2: Unnest the tibble

Tibbles are new data frames that keep the features we like and drops the features that are now frustrating (i.e. converting character vectors to factors). - unnest() the tibble - remove the lines that are LaTeX crude - compute a line number with the mutate function

```
quantum_words <- quantum_words %>%
   unnest() %>%
   mutate(line_number = 1:n(),
      file =
      str_sub(basename(file), 1, -5))
```

Step 3: Delete words and LaTex

- str_detect() detects the presence or absence of a pattern and returns a logical vector
- ! is the logical operator for negation

2.14.2 Quantum Words

```
quantum_words <- quantum_words %>%
  unnest() %>%
  filter(text != "%!TEX root = ind.tex") %>%
  filter(!str_detect(text, "^(\\\\[A-Z,a-z])"), text != "")
```

2.14.3 Quantum Tokens

Here, we tokenize the Quantum corpus as words and shw the most frequent words shared by the three texts. However, we have done this without removing common words that do not contribute to the meaning of the corpus, like "I," and "do."

```
quantum_words <- quantum_words %>%
  unnest_tokens(word, text) %>%
  filter(!str_detect(word, "[0-9]"),
         word != "fismanreview",
         word != "multicolumn",
         word != "p",
         word != "_i",
         word != "al",
         word != "tabular",
         word != "ref",
         word != "cite",
         !str_detect(word, "[a-z]_"),
         !str_detect(word, ":"),
         word != "bar",
         word != "emph",
         !str_detect(word, "textless"))
quantum_words
```

```
# A tibble: 158,482 x 3
   file           line_number word
   <chr>                <int> <chr>
 1 quantum_phaith           1 quantum
 2 quantum_phaith           1 phaith
 3 quantum_phaith           2 preface
 4 quantum_phaith           3 obviously
 5 quantum_phaith           3 either
 6 quantum_phaith           3 i
 7 quantum_phaith           3 do
 8 quantum_phaith           3 not
 9 quantum_phaith           3 know
10 quantum_phaith           3 how
# ... with 158,472 more rows
```

2.14.4 Quantum Cloud

We represent the frequently occurring words in the Quantum corpus with the word cloud shown in Figure 2-7, and the code appears here as well.

```
library(reshape2)
quantum_words %>%
  inner_join(get_sentiments("bing")) %>%
  count(word, sentiment, sort = TRUE) %>%
  acast(word ~ sentiment, value.var = "n", fill = 0) %>%
```

```
comparison.cloud(colors = c("red", "blue"),
                 max.words = 100)
```

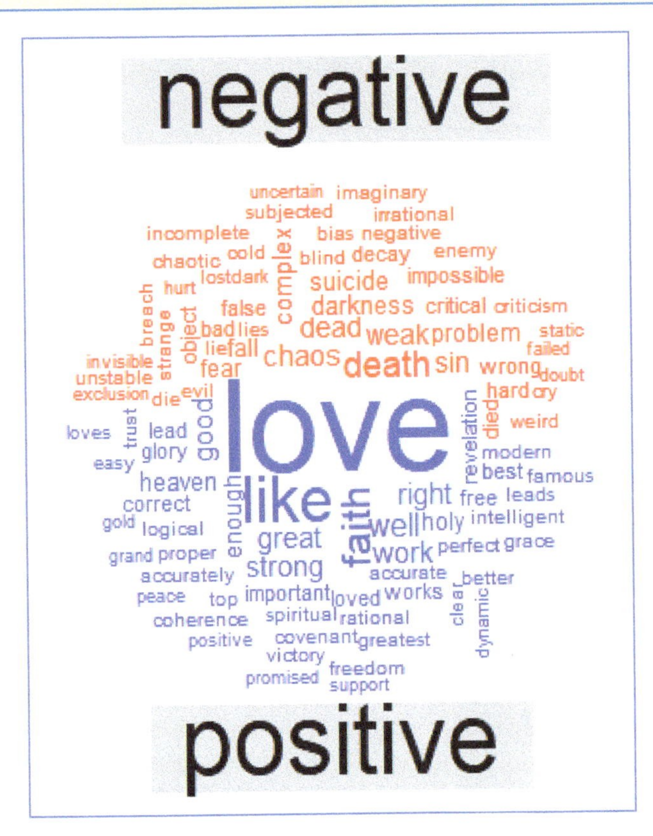

Figure 2-7. Word cloud representation of the Quantum corpus.

2.14.5 Lexicon Exploration

The *tidytext* package contains several sentiment lexicons in the sentiments dataset. The three general-purpose lexicons are:

- `AFINN` from Finn Årup Nielsen
- `Bing` from Bing Liu and collaborators
- `Loughran` from Tim Loughran and Bill McDonald

All three of these lexicons are based on unigrams, i.e., single words.

```
library(tidytext)
sentiments
```

```
# A tibble: 27,314 x 4
   word        sentiment lexicon  score
   <chr>       <chr>     <chr>    <int>
 1 abacus      trust     loughran    NA
 2 abandon     fear      loughran    NA
 3 abandon     negative  loughran    NA
 4 abandon     sadness   loughran    NA
 5 abandoned   anger     loughran    NA
 6 abandoned   fear      loughran    NA
 7 abandoned   negative  loughran    NA
 8 abandoned   sadness   loughran    NA
 9 abandonment anger     loughran    NA
10 abandonment fear      loughran    NA
# ... with 27,304 more rows
```

2.14.6 Loughran Lexicon

These lexicons contain many English words and the words are assigned scores for positive/negative sentiment, and also possibly emotions like joy, anger, sadness, and so forth. The `loughran` lexicon: - Includes 13,901 words - Categorizes words in a binary fashion ("yes"/"no") into categories of: - Positive - Negative - Anger - Fear - Joy - Sadness - Surprise - Trust

```
get_sentiments("loughran")
```

```
# A tibble: 13,901 x 2
   word        sentiment
   <chr>       <chr>
 1 abacus      trust
 2 abandon     fear
 3 abandon     negative
 4 abandon     sadness
 5 abandoned   anger
 6 abandoned   fear
 7 abandoned   negative
 8 abandoned   sadness
 9 abandonment anger
10 abandonment fear
# ... with 13,891 more rows
```

2.14.7 AFINN Lexicon

- Includes 2,476 words
- The `AFINN` lexicon assigns words with a score that runs between -5 and 5 with
  - negative scores indicating negative sentiment
  - positive scores indicating positive sentiment.

```
get_sentiments("afinn")
```

```
# A tibble: 2,476 x 2
   word       score
   <chr>      <int>
 1 abandon       -2
 2 abandoned     -2
 3 abandons      -2
 4 abducted      -2
 5 abduction     -2
 6 abductions    -2
 7 abhor         -3
 8 abhorred      -3
 9 abhorrent     -3
10 abhors        -3
# ... with 2,466 more rows
```

2.14.8 Bing Lexicon

- Includes 13,901 words
- The bing lexicon categorizes words in a binary fashion ("yes"/"no") into categories of:
  - Positive
  - Negative
  - Anger
  - Fear
  - Joy
  - Sadness
  - Surprise
  - Trust

```
get_sentiments("bing")
```

```
# A tibble: 6,788 x 2
   word        sentiment
   <chr>       <chr>
 1 2-faced     negative
 2 2-faces     negative
 3 a+          positive
 4 abnormal    negative
 5 abolish     negative
 6 abominable  negative
 7 abominably  negative
 8 abominate   negative
 9 abomination negative
10 abort       negative
# ... with 6,778 more rows
```

2.14.9 Getting Sentiments with Loughran

Using the Loughran lexicon, let's see how the emotions of my words change between Quantum Phaith and Quantum Love.

```
quantum_words %>%
  inner_join(get_sentiments("loughran")) %>%
  group_by(index = line_number %/% 25, file, sentiment) %>%
  summarize(n = n()) %>%
  ggplot(aes(x = index, y = n, fill = file)) +
  geom_bar(stat = "identity", alpha = 0.8) +
  facet_wrap(~ sentiment, ncol = 5)
```

The plot appears on the following page.

Based on the plot, it looks like Quantum Love is mode "sentimental, as the sentiment scores are a little higher than for Quantum Phaith. Not that Quantum Phaith is much more scientific.

2.14.10 Analyzing word and document frequency: tf-idf

A central question in text mining and natural language processing (NLP) is how to quantify what a document is about. Can we do this by looking at the words that make up the document? One measure of how important a word may be is its *term frequency* (tf), how frequently a word occurs in a document, as we examined in Chapter 1. There are words in a document, however, that occur many times but may not be important; in English, these are probably words like "the", "is", "of", and so forth. We might take the approach of adding words like these to a list of stop words and removing them before analysis, but it is possible that some of these words might be more important in some documents than others.

Definition 2.8. *Stopwords are the words in any language which does not add much meaning to a sentence.*

A list of stop words is not a very sophisticated approach to adjusting term frequency for commonly used words. Moreover, there is not a universal list of stop words. They can best be considered as contextual. That is, in some documents they add value and in ortehr they do not. Figure 2-8 shows the most frequent words shared by the three texts after the removal of stopwords.

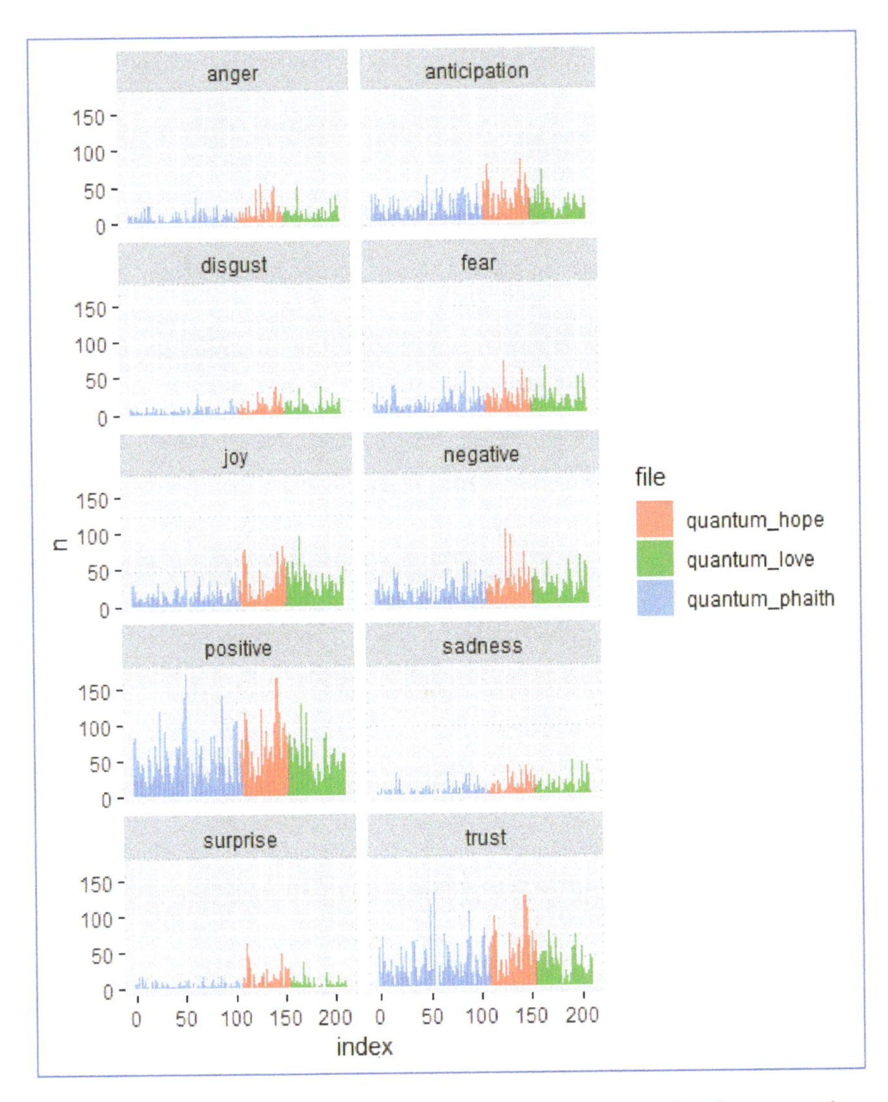

Figure 2-8. Histograms for morst frequent Quantum words after removing stopwords.

Another approach is to look at a term's **inverse document frequency** (idf), which decreases the weight for commonly used words and increases the weight for words that are not used very much in a collection of documents. This can be combined with **term frequency** to calculate a term's **tf-idf** (the two quantities multiplied together), the frequency of a term adjusted for how rarely it is used.

The statistic tf-idf is intended to measure how important a word is to a document in a collection (or corpus) of documents, for example, to one novel in a collection of novels or to one website in a collection of websites.

2.14.11 Term frequency

Let's start by looking at a corpus of four books and examine first term frequency, then tf-idf. We can start just by using *dplyr* verbs such as `group_by()` and `join()`. What are the most commonly used words in our book corpus? (Let's also calculate the total words in each novel here, for later use.)

First, we load and tokenize the book corpus called: readings," as well as remove stopwords.

```
readings <- read.csv("C:\\Users\\jeff\\Documents\\VIT_Course_Mat
erial\\Data_Analytics_2018\\data\\readings.csv",stringsAsFactor=
FALSE)

readings<-readings %>% group_by(book) %>% mutate(ln=row_number()
)%>% unnest_tokens(word,text) %>% count(book, word, sort = TRUE)
%>%ungroup()

my_stops <- c("https", "http", stopwords("en"))
readings <- readings %>%
  filter(!word %in% stop_words$word,
         !word %in% my_stops,
         !word %in% str_remove_all(stop_words$word, "'"))
head(readings,5)
```

```
# A tibble: 5 x 3
  book              word        n
  <chr>             <chr>   <int>
1 Indian Philosophy world    1073
2 Indian Philosophy life      693
3 Indian Philosophy god       660
4 Quantum Love      love      604
5 Indian Philosophy buddha    566
```

2.14.12 Get Term Frequencies

So far, we have taken the corpus of Quantum texts and put them all in one document, for all practical purposes. Now, we want to exlore the corpus by **joining** the three Quantum texts and the two Indian philosophy texts. Then we calculate the word frequencies for shared words.

```
total_words <- readings %>%
  group_by(book) %>%
  summarize(total = sum(n))

book_words <- left_join(readings, total_words)
 Joining, by = "book"
book_words
```

```
# A tibble: 31,228 x 4
   book              word           n total
   <chr>             <chr>      <int> <int>
 1 Indian Philosophy world       1073 93145
 2 Indian Philosophy life         693 93145
 3 Indian Philosophy god          660 93145
 4 Quantum Love      love         604 12077
 5 Indian Philosophy buddha       566 93145
 6 Indian Philosophy reality      538 93145
 7 Indian Philosophy nature       451 93145
 8 Indian Philosophy soul         443 93145
 9 Indian Philosophy upanisads    439 93145
10 Indian Philosophy existence    437 93145
# ... with 31,218 more rows
```

There is one row in this book_words data frame for each word-book combination; n is the number of times that word is used in that book and total is the total words in that book. The usual suspects are here with the highest n, "the", "and", "to", and so forth. The plot shows the distribution of n/total for each novel, the number of times a word appears in a novel divided by the total number of terms (words) in that book. This is exactly what "term frequency" is.

```
library(ggplot2)
ggplot(book_words, aes(n/total, fill = book)) +
  geom_histogram(show.legend = FALSE) +
  xlim(NA, 0.0009) +
  facet_wrap(~book, ncol = 2, scales = "free_y")
```

```
`stat_bin()` using `bins = 30`. Pick better value with `binwidth
`.
Warning: Removed 584 rows containing non-finite values (stat_bin
).
```

The plot appears as on the following page.

2.14.13 Zipf's law

Distributions like those shown in the previous plot are typical in language. In fact, those types of long-tailed distributions are so common in any given corpus of natural language (like a book, or a lot of text from a website, or spoken words) that the relationship between the frequency that a word is used and its rank has been the subject of study; a classic version of this relationship is called Zipf's law, after George Zipf, a 20th century American linguist. This is shown in Figure 2-9.

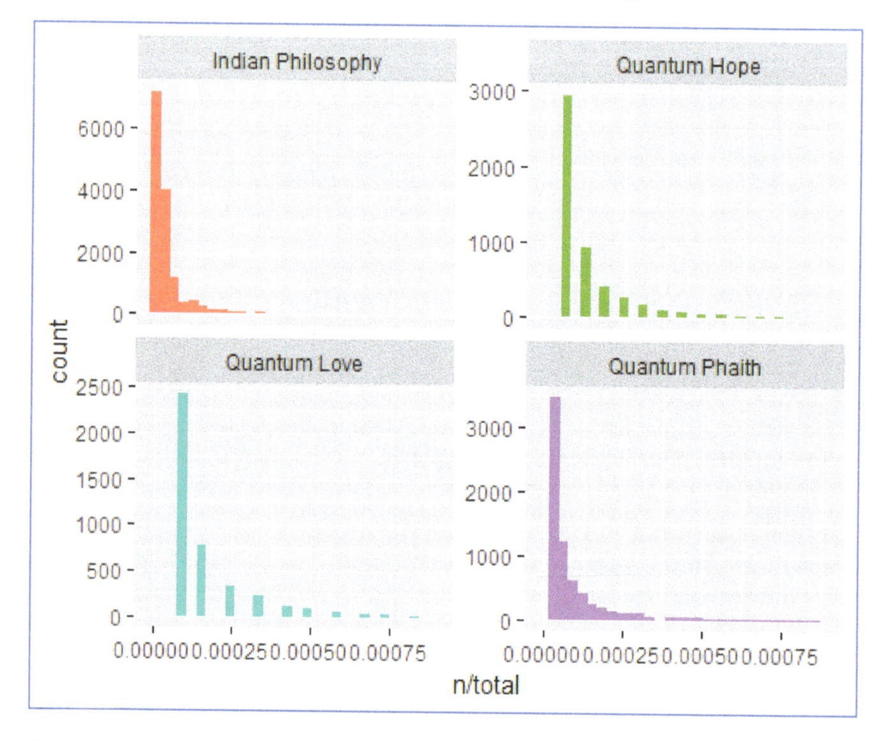

Figure 2-9. Histograms of the four texts using Zipf's law.

Zipf's law states that the frequency that a word appears is inversely proportional to its rank.

> ***Definition 2.9.*** *The probability of occurrence of words or other items starts high and tapers off. Thus, a few occur very often while many others occur rarely.*

Since we have the data frame, we used to plot term frequency, we can examine Zipf's law for our book collect with just a few lines of dplyr functions. By doing this, we can compute the rank of the frequent words, using Zipf's law.

```
freq_by_rank <- book_words %>%
  group_by(book) %>%
  mutate(rank = row_number(),
         `term frequency` = n/total)
freq_by_rank
```

```
# A tibble: 31,228 x 6
# Groups:   book [4]
   book              word          n total  rank `term frequency`
   <chr>             <chr>     <int> <int> <int>            <dbl>
 1 Indian Philosophy world      1073 93145     1          0.0115
 2 Indian Philosophy life        693 93145     2          0.00744
 3 Indian Philosophy god         660 93145     3          0.00709
 4 Quantum Love      love        604 12077     1          0.0500
 5 Indian Philosophy buddha      566 93145     4          0.00608
 6 Indian Philosophy reality     538 93145     5          0.00578
 7 Indian Philosophy nature      451 93145     6          0.00484
 8 Indian Philosophy soul        443 93145     7          0.00476
 9 Indian Philosophy upanisads   439 93145     8          0.00471
10 Indian Philosophy existence   437 93145     9          0.00469
# ... with 31,218 more rows
```

The rank column here tells us the rank of each word within the frequency table; the table was already ordered by n so we could use row_number() to find the rank. Then, we can calculate the term frequency in the same way we did before. Zipf's law is often visualized by plotting rank on the x-axis and term frequency on the y-axis, on logarithmic scales. Plotting this way, an inversely proportional relationship will have a constant, negative slope.

```
freq_by_rank %>%
  ggplot(aes(rank, `term frequency`, color = book)) +
  geom_line(size = 1.1, alpha = 0.8, show.legend = FALSE) +
  scale_x_log10() +
  scale_y_log10()
```

The plot appears as Figure 2-10 following page.

Notice that the plot is in log-log coordinates. We see that all four of our books are similar to each other, and that the relationship between rank and frequency does have negative slope. It is not quite constant, though; perhaps we could view this as a broken power law with, say, three sections. Let's see what the exponent of the power law is for the middle section of the rank range.

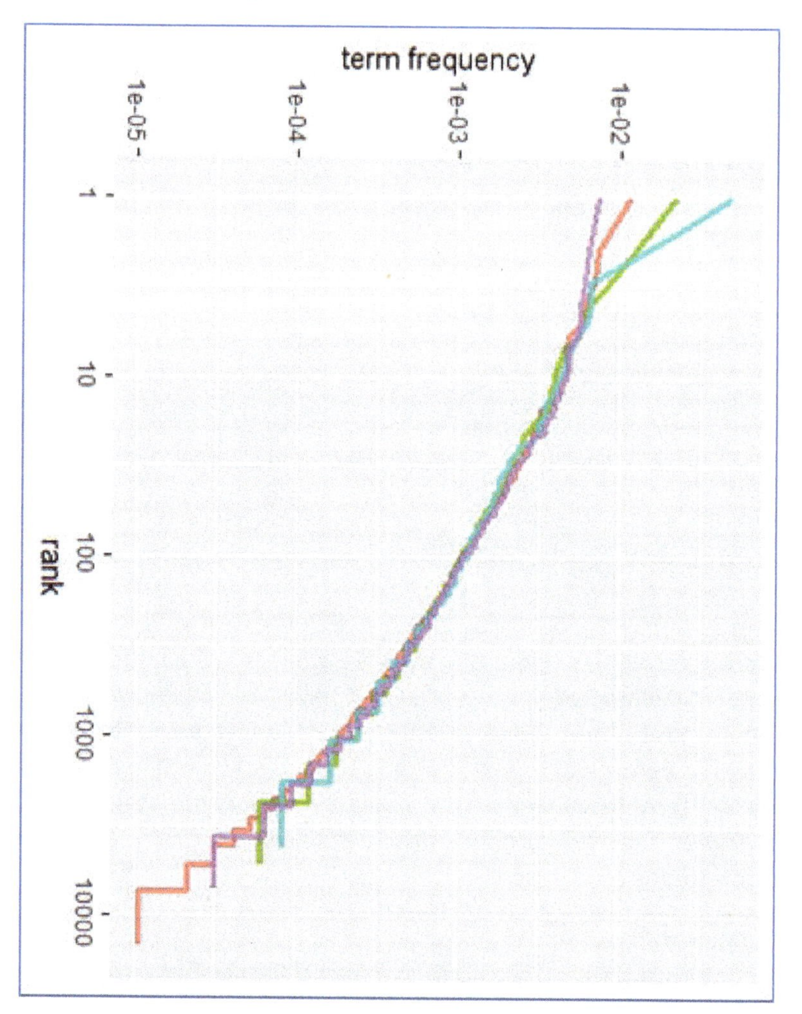

Figure 2-10. Zipf's law is often visualized by plotting rank on the x-axis and term frequency on the y-axis, on logarithmic scales.

```
rank_subset <- freq_by_rank %>%
  filter(rank < 500,
         rank > 10)
lm(log10(`term frequency`) ~ log10(rank), data = rank_subset)
```

```
Call:
 lm(formula = log10(`term frequency`) ~ log10(rank), data = rank
_subset)
```

```
Coefficients:
(Intercept)  log10(rank)
    -1.6528      -0.6521
```

Classic versions of Zipf's law have

$$frequency \propto \frac{1}{rank}$$

and we have in fact gotten a slope close to -1 here. Let's plot this fitted power law with the data in figure below to see how it looks

```
freq_by_rank %>%
  ggplot(aes(rank, `term frequency`, color = book)) +
  geom_abline(intercept = -0.62, slope = -1.1, color = "gray50",
linetype = 2) +
  geom_line(size = 1.1, alpha = 0.8, show.legend = FALSE) +
  scale_x_log10() +
  scale_y_log10()
```

The plot appears in Figure 2-11 on the following page.

We have found a result close to the classic version of Zipf's law for the corpus of four "philosophy" books. The deviations we see here at high rank are not uncommon for many kinds of language; a corpus of language often contains fewer rare words than predicted by a single power law. The deviations at low rank are more unusual. Our books use a lower percentage of the most common words than many collections of language. This kind of analysis could be extended to compare authors, or to compare any other collections of text; it can be implemented simply using tidy data principles.

2.14.14 The bind_tf_idf function

The idea of tf-idf is to find the important words for the content of each document by decreasing the weight for commonly used words and increasing the weight for words that are not used very much in a collection or corpus of documents, in this case, the group of books as a whole. Calculating tf-idf attempts to find the words that are important (i.e., common) in a text, but not too common. Let's do that now.

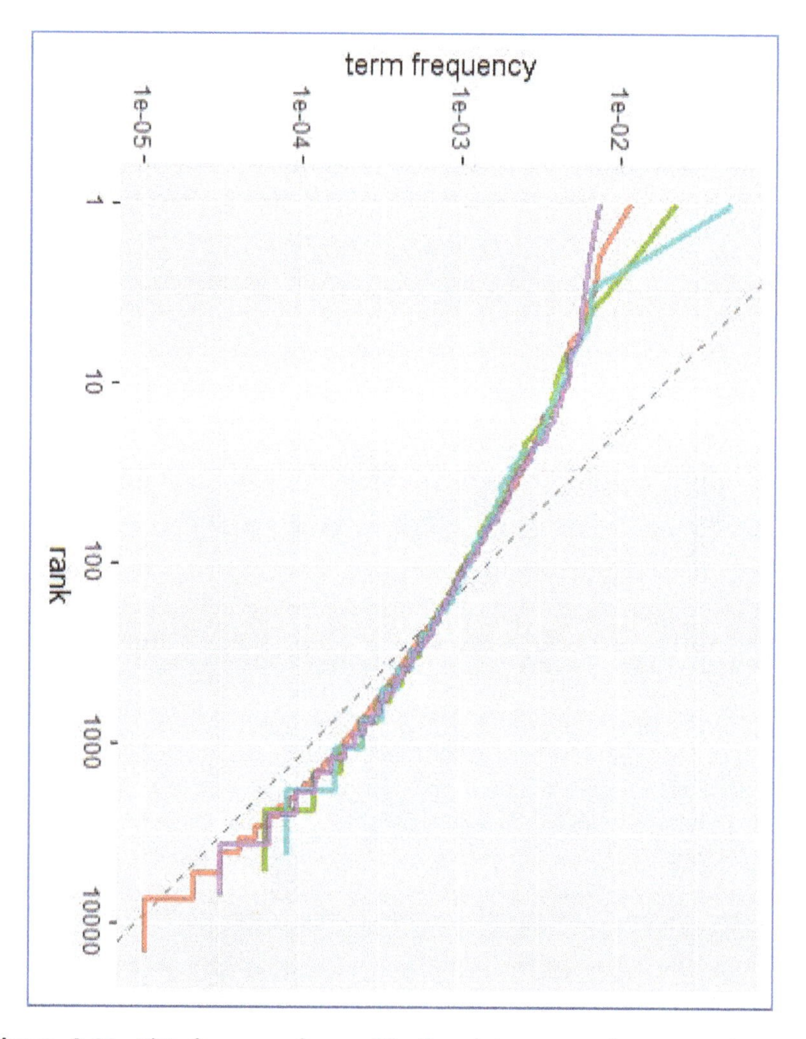

Figure 2-11. Fitted power law with the data appearing step-wize and interpolation is a dashed line.

The `bind_tf_idf()` function in the tidytext package takes a tidy text dataset as input with one row per token (term), per document. One column (word here) contains the terms/tokens, one column contains the documents (book in this case), and the last necessary column contains the counts, how many times each document contains each term (n in this example). We calculated a total for each book for our explorations in previous sections, but it is not necessary for the `bind_tf_idf()` function; the table only needs to contain all the words in each document.

```r
book_words <- book_words %>%
  bind_tf_idf(word, book, n)
book_words
```

```
# A tibble: 31,228 x 7
   book               word          n total      tf   idf  tf_idf
   <chr>              <chr>     <int> <int>   <dbl> <dbl>   <dbl>
 1 Indian Philosophy world      1073 93145 0.0115   0     0
 2 Indian Philosophy life        693 93145 0.00744  0     0
 3 Indian Philosophy god         660 93145 0.00709  0     0
 4 Quantum Love      love        604 12077 0.0500   0     0
 5 Indian Philosophy buddha      566 93145 0.00608  1.39 0.00842
 6 Indian Philosophy reality     538 93145 0.00578  0     0
 7 Indian Philosophy nature      451 93145 0.00484  0     0
 8 Indian Philosophy soul        443 93145 0.00476  0     0
 9 Indian Philosophy upanisads   439 93145 0.00471  1.39 0.00653
10 Indian Philosophy existence   437 93145 0.00469  0     0
# ... with 31,218 more rows
```

Notice that idf and thus tf-idf are zero for these extremely common words. These are all words that appear in all all four books, so the idf term (which will then be the natural log of 1) is zero. The inverse document frequency (and thus tf-idf) is very low (near zero) for words that occur in many of the documents in a collection; this is how this approach decreases the weight for common words. The inverse document frequency will be a higher number for words that occur in fewer of the documents in the collection.

Let's look at terms with high tf-idf in book collection.

```r
book_words %>%
  select(-total) %>%
  arrange(desc(tf_idf))
```

```
# A tibble: 31,228 x 6
   book              word              n      tf   idf  tf_idf
   <chr>             <chr>         <int>   <dbl> <dbl>   <dbl>
 1 Indian Philosophy buddha          566 0.00608 1.39  0.00842
 2 Indian Philosophy upanisads       439 0.00471 1.39  0.00653
 3 Indian Philosophy buddhism        313 0.00336 1.39  0.00466
 4 Indian Philosophy brahman         260 0.00279 1.39  0.00387
 5 Quantum Hope      quantum         205 0.0126  0.288 0.00362
 6 Quantum Phaith    calendar        134 0.00429 0.693 0.00297
 7 Indian Philosophy veda            186 0.00200 1.39  0.00277
 8 Indian Philosophy nirvana         180 0.00193 1.39  0.00268
 9 Indian Philosophy consciousness   344 0.00369 0.693 0.00256
10 Quantum Phaith    elohim           54 0.00173 1.39  0.00240
# ... with 31,218 more rows
```

Here we see all proper nouns, names that are in fact important in these novels. None of them occur in all of novels, and they are important, characteristic words for each text within the corpus of books.

We wil examine a visualization for these high tf-idf words in the next plot. This is shown in Figure 2-12.

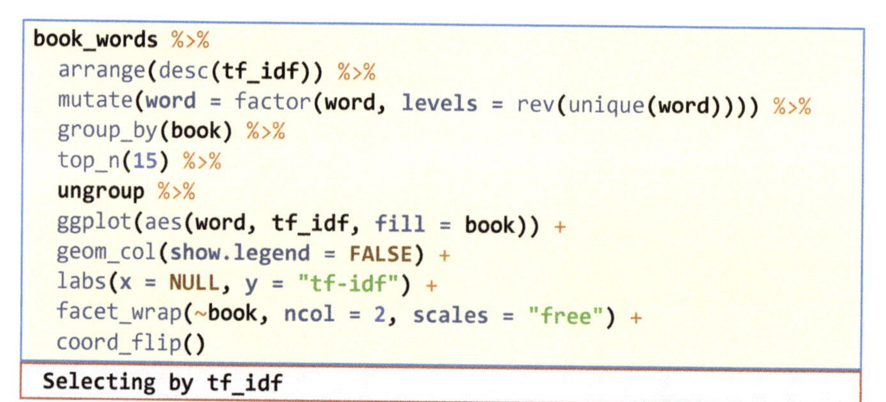

```
book_words %>%
  arrange(desc(tf_idf)) %>%
  mutate(word = factor(word, levels = rev(unique(word)))) %>%
  group_by(book) %>%
  top_n(15) %>%
  ungroup %>%
  ggplot(aes(word, tf_idf, fill = book)) +
  geom_col(show.legend = FALSE) +
  labs(x = NULL, y = "tf-idf") +
  facet_wrap(~book, ncol = 2, scales = "free") +
  coord_flip()
```

```
Selecting by tf_idf
```

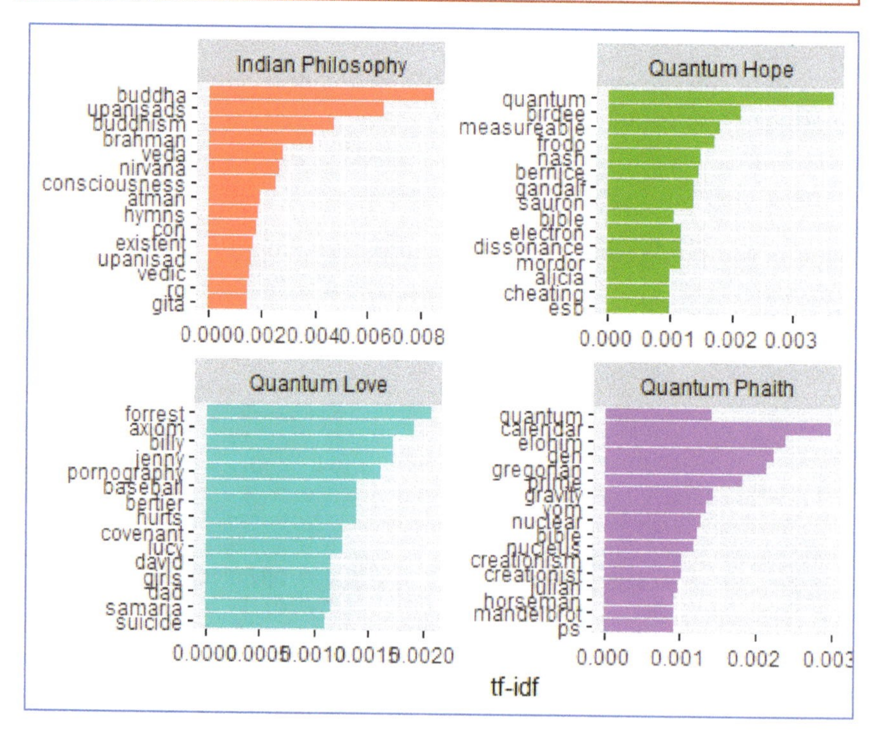

Figure 2-12. *Histogram visualization for the high tf-idf words.*

We still see all proper nouns! These words are, as measured by tf-idf, the most important to each novel and most readers would likely agree. What measuring tf-idf has done here is show us that our authors used similar language across their four books. This is the point of tf-idf; it identifies words that are important to one document within a collection of documents.

2.14.15 Summary

Using term frequency and inverse document frequency allows us to find words that are characteristic for one document within a collection of documents, whether that document is a novel or physics text or webpage. Exploring term frequency on its own can give us insight into how language is used in a collection of natural language, and *dplyr* verbs like `count()` and `rank()` give us tools to reason about term frequency. The tidytext package uses an implementation of tf-idf consistent with tidy data principles that enables us to see how different words are important in documents within a collection or corpus of documents.

Markdown Note		
Results Example: ```{r echo = TRUE}		
Collapse	FALSE	If TRUE, knitr will collapse all the source and output blocks created by the chunk into a single block.
Echo	TRUE	If FALSE, knitr will not display the code in the code chunk above it's results in the final document.
Results	'markup'	If 'hide', knitr will not display the code's results in the final document. If 'hold', knitr will delay displaying all output pieces until the end of the chunk. If 'asis', knitr will pass through results without reformatting them (useful if results return raw HTML, etc.)
Message	TRUE	If FALSE, knitr will not display any messages generated by the code.
Warning	TRUE	If FALSE, knitr will not display any warning messages generated by the code.

2.15 Exercises

1. Choose and download five text files from *https://github.com/stricje1/VIT_University/tree/master/Predictive_Modeling/tidy_text*. ***And put I into a corpus (directory) of your local drive.***

a. Assign a file path to the variable cname, using your local drive corpus directory file.path("C:/Users/<your name>/<subdirectory name>/.../data", "text_files").
b. Create a corpus, i.e. `docs <- Corpus(DirSource(cname))`
c. Inspect the corpus and record the metadata.
d. Convert the text to lowercase and inspect your work.
e. Remove unnecessary words from the text.
f. If appropriate, combine words that should stay together.
g. Put the text into a *term-document matrix* and inspect it.
h. Create *document-term matrix* and inspect it.
i. Organize the terms by their frequency and put it into a matrix and save it to your working directory.
j. Remove sparse terms.
k. Check out some of the most and least frequently occurring words.
l. Build a frequency table.
m. Plot a frequency distribution of the most frequent terms.
n. Find correlations in the text.
o. Plot words using a wordcloud that occur at least X (determine X from your frequency table) times.

2. Pick your favorite song and put it into an array, **c**(), like we did earlier in the chapter.
   a. Print out the tibble in R using the dplyr and tibble packages.
   b. Break the text into individual tokens using tidytext's unnest_tokens() function.
   c. What are the resulting word frequencies?

3. Use the collection of Associated Press newspaper articles included in the topicmodels package and analyze the data with tidy tools. [Hint: use `data("AssociatedPress", package = "topicmodels")`.]
   a. How many documents and terms are in the data set?
   b. What the sparsity of the data?
   c. Turn the text into a data frame with one-token-per-document-per-row.

d. Perform sentiment analysis on these newspaper articles with the approach described in this chapter.
e. Use tables and graphs to present your results.

Chapter 3 – Vectorizing Text Data

3.1 Introduction

In simplest terms, **word embeddings** are the texts converted into numbers. Given that, there may be different numerical representations of the same text. However, before we start examining the details of word embeddings, we should ask: Why do we need word embeddings?

Many machine learning (ML) algorithms and nearly all Deep Learning Architectures (DLAs) are incapable of processing strings or plain text in their raw form. They require numbers as inputs to perform any sort of analysis, be it classification, regression, and so on. With the tremendous amount of data available in the text format, it is paramount to extract information out of the data and build appropriate applications. Some real-world applications of text applications are sentiment analysis of reviews performed by Amazon, and news classification (or clustering) performed by Google.

3.2 Definitions

A **word embedding** is format that generally attempts to map a word, using a dictionary, to a numeric vector.

> ***Definition 3.1.*** *A word embedding is a learned representation for text where words that have the same meaning have a similar representation.*

For example, we can break down the following sentence into words as tokens.

"These are the best of times."

A word in this sentence may be treated as a token and a dictionary may be a list of all tokens (unique words) in the sentence:

['are', 'best', 'of', 'the', 'these' , 'times']

A vector representation of a word may be a binary encoded vector where 1 stands for the position where the word exists and 0 everywhere else. The vector representation of "best" format, according to the above dictionary, is [0,1,0,0,0,0] and the sentence converted as [0,0,0,1,0,0], where 'best' is the fourth token in the sentence. This is just a very simple method to represent a word in the vector form.

__Definition 3.2.__ Vectorization is the process of converting an algorithm from operating on a single value at a time to operating on a set of values (vector) at one time.

We now turn to examine different types of word embeddings or **word vectors** and their advantages and disadvantages over the rest.

3.3 Types of Word Embeddings

The different types of word embeddings can be broadly classified into two categories:

- Frequency-based Embedding
- Prediction-based Embedding

Let us try to understand each of these methods in detail.

3.3.1 Frequency-based Embedding

Let's examine two types of frequency-based embeddings (there are more), the **count vector** and the **TF-IDF vector**.

Count Vector

An encoded vector is returned with a length of the entire vocabulary and an integer count for the number of times each word appeared in the document.

Definition 3.3. *Consider a Corpus C comprised of D documents $\{d_1, d_2 \ldots\ldots d_D\}$ and N unique tokens extracted out of the corpus C. The N tokens will form our dictionary and the size of the Count Vector matrix M will be $D \times N$. Each row in the matrix M contains the frequency of tokens in document D_i.*

Let us look at this using a simple example, each document is a sentence and their tokens are words.

D_1: "These are the best of times"

D_2: "These are the worst of times"

The dictionary created may be a list of unique tokens(words) in the corpus = ['are', 'best', 'of', 'the', 'these' , 'times', 'worst']. In this instance, $D = 2, N = 6$.

The count matrix M of size 2×6 will be represented as :

	are	best	of	the	these	times	worst
$\boldsymbol{D_1}$	1	1	1	1	1	1	0
$\boldsymbol{D_2}$	1	0	1	1	1	1	1

Now, a column can also be understood as word vector for the corresponding word in the matrix M. For example, the word vector for 'best' in the above matrix is [1,0] and so on. Here, the rows correspond to the documents in the corpus and the columns correspond to the tokens in the dictionary. The second row in the above matrix may be read as D_2 contains all words, except 'best', once.

Now there may be quite a few variations while preparing the above matrix M. The variations will be generally in:

The way the dictionary is prepared. Why? Because in real world applications we might have a corpus which contains millions of documents. And with millions of documents, we can extract hundreds of millions of unique words. So basically, the matrix that will be prepared like above will be a very sparse one and inefficient for any computation. So, an alternative to using every unique word as a dictionary element would be to pick say top 10,000 words based on frequency and then prepare a dictionary.

The way the count is taken for each word. We may either take the frequency (number of times a word appears in the document) or the presence (has the word appeared in the document? [yes=1/no=0]) to be the entry in the count matrix M. But generally, frequency method is preferred over the latter.

Let's add two more documents to our corpus:

D_3: "This is the age of wisdom"

D_4: "This is the age of foolishness"

Then our count matrix M of size 2×6 will be represented as :

	age	are	best	foolishness	is	of	the	these	this	times	wisdom	worst
D_1	0	1	1	0	0	1	1	1	0	1	0	0
D_2	0	1	0	0	0	1	1	1	0	1	0	1
D_3	1	0	0	0	1	1	1	0	1	0	1	0
D_4	1	0	0	1	1	1	1	0	1	0	0	0

For convienence, we can transpose this matrix M as M^T:

	D_1	D_2	D_3	D_4
age	0	0	1	1
are	1	1	0	0
best	1	0	0	0
foolishness	0	0	0	1
is	0	0	1	1
of	1	1	1	1
the	1	1	1	1
these	1	1	0	0
this	0	0	1	1
times	1	1	0	0
wisdom	0	0	1	0
worst	0	1	0	0

Then, the word vector for 'age' is $[0, 0, 1, 1]$. Also notice that the word vectors are not unique, for instance, 'of' and 'the' are both $[1, 1, 1, 1]$.

TF-IDF vectorization

Term frequency times inverse document frequency (tf-idf) is another word embedding method, which is based on the frequency method. However, it is different from the count vectorization in the sense that it takes into account a single word in the entire corpus rather than the occurrence of a word in a single document.

Definition 3.4. *Consider a Corpus C comprised of D documents $\{d_1, d_2 \ldots\ldots d_D\}$ and N unique tokens extracted out of the corpus C. The tf-idf vector measures the importance of a token by comparing it to the frequency of the token in a large set of documents.*

Common words like 'a', 'the', 'is', etc. tend to appear quite frequently in documents in comparison to the words which are important to a document. For example, a document about the 1863 Battle of Gettysburg will contain more occurences of the word "Chamberlain"[2] in comparison to other United States Civil War documents. But common words like "the" or perhaps "union" are also going to be present in higher frequency in almost every Civil War document.

Ideally, what we would want is to penalize the common words occurring in almost all documents and give more importance to words that appear in a subset of documents. TF-IDF works for these common words by assigning them lower weights, while giving importance to words like Chamberlain in a particular document.

Term	Count
this	4
is	2
about	1
Chamberlain	4
and	1
Gettysburg	2

Term	Count
this	3
is	2
about	1
Sherman	2
and	1
Atlanta	3

2/7 and 1/3

Let's consider to use TF-IDF to discover important terms in these two documents.

---

[2] Joshua Lawrence Chamberlain was awarded the Congressional Medal of Honor for his part in the defense of Little Roundtop by the 20th Maine, a Union infantry regiment, during the Battle of Gettysburg in 1863.

Term Frequency (TF) is a scoring of the frequency of the word in the current document. Since every document is different in length, it is possible that a term would appear much more times in long documents than shorter ones. The term frequency is often divided by the document length to normalize.

$$TF(t) = \frac{number\ of\ times\ term\ t\ appears\ in\ a\ document}{Total\ number\ of\ terms\ in\ the\ document}$$

So,

$$TF(this, Doument\ 1) = \frac{4}{14} = \frac{2}{7}$$

and

$$TF(this, Document\ 2) = \frac{3}{12} = \frac{1}{4},$$

while

$$TF(Chamberlain, Document\ 1) = \frac{4}{14} = \frac{2}{7}$$

Inverse Document Frequency (IDF) is a scoring of how rare the word is across documents. It is a measure of how rare a term is. So that the rarer the term, the higher the IDF score is.

$$IDF(t) = \ln\left(\frac{Total\ number\ of\ documents}{Number\ of\ documents\ containing\ term\ t}\right)$$

So,

$$IDF(\text{this}) = \ln\left(\frac{2}{2}\right) = \ln(1) = 0$$

Now, let's calculate the value of "Chamberlain":

$$IDF(\text{this}) = \ln\left(\frac{2}{1}\right) = \ln(2) = 0.301$$

TF-IDF weight is a statistical measure used to evaluate how important a word is to a document in a collection or corpus. The importance increases proportionally to the number of times a word appears in the document but is offset by the frequency of the word in the corpus. So,

The $TF - IDF\ score = TF * IDF$, and

$$TF - IDF(this, Document\ 1) = \left(\frac{1}{36}\right) \times (0) = 0$$

and

$$TF - IDF(Chamberlain, Document\ 1) = \left(\frac{2}{7}\right) \times (0.301) = 0.86$$

So, in Document 1 both 'this' and 'Chamberlain' have the same frequency, but 'this' is penalized by IDF for its appearance in both doucuments, resulting in 'Chamberlain' being weighted by its unique appearance in Document 1.

3.3.2 Prediction-based Embedding

The second type of word embeddings is called the prediction-based vector or commonly called Word2Vec, and is comprised of the **skip-gram** and **continuous bag of words algorithms (cbow)**. These are based on an understanding of neual networks and are beyond the scope of this text. For an explanantion, see Mikolav, et al (Mikolov, Le, & Sutskever, 2013).

3.4 Plain Text Technical Details

It is worth noting some technical requirements when we work with plain text data. If you use a word processor to edit text or you are analyzing test that someone else wrote, in Microsoft Word, for example,the text is probally rich text format and has punctuation like culry quotes ("_"). Neither Python or R understands curly quotes—you tect must have straith quotes, like "fourscore" and 'we' (as in the text below). A good work-around is a good text editor. We use Notepad++, having tried many others. Notepad++ is free and easy to use. It allows multiple documents (window) to be open and save them as temp files even if you do not think to do it, so that your documents will be available when you reopen the app. It also saves multiple formats, like text (*.txt), Python (*.py), R (*.r), and SQL (*.sql). Other formats are shown in Figure 3-1.

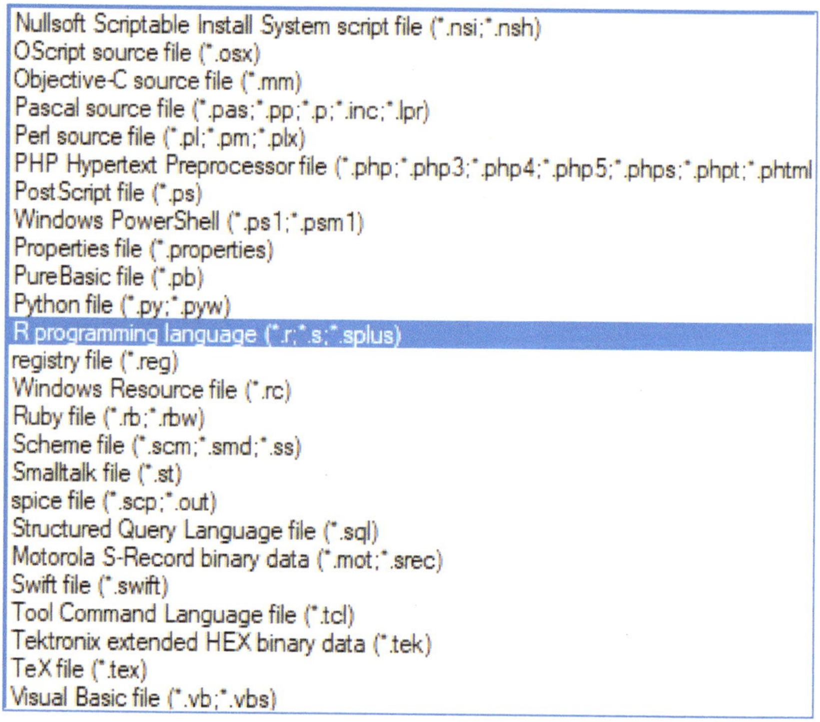

Nullsoft Scriptable Install System script file (*.nsi;*.nsh)
OScript source file (*.osx)
Objective-C source file (*.mm)
Pascal source file (*.pas;*.pp;*.p;*.inc;*.lpr)
Perl source file (*.pl;*.pm;*.plx)
PHP Hypertext Preprocessor file (*.php;*.php3;*.php4;*.php5;*.phps;*.phpt;*.phtml
PostScript file (*.ps)
Windows PowerShell (*.ps1;*.psm1)
Properties file (*.properties)
PureBasic file (*.pb)
Python file (*.py;*.pyw)
R programming language (*.r;*.s;*.splus)
registry file (*.reg)
Windows Resource file (*.rc)
Ruby file (*.rb;*.rbw)
Scheme file (*.scm;*.smd;*.ss)
Smalltalk file (*.st)
spice file (*.scp;*.out)
Structured Query Language file (*.sql)
Motorola S-Record binary data (*.mot;*.srec)
Swift file (*.swift)
Tool Command Language file (*.tcl)
Tektronix extended HEX binary data (*.tek)
TeX file (*.tex)
Visual Basic file (*.vb;*.vbs)

Figure 3-1. *Notepad++ file format options (there are other not shown here).*

3.5 Word Vectorizing with Python

Now, let's turn to an example of frequency-based embeddings using Python. For the example, we use the Gettysburg Address by Abraham Lincoln in 1863[3].

```
text = "Fourscore and seven years ago our fathers brought forth
on this continent a new nation, conceived in liberty and dedicat
ed to the proposition that all men are created equal. Now we are
engaged in a great civil war, testing whether that nation or any
nation so conceived and so dedicated can long endure. We are met
on a great battlefield of that war. We have come to dedicate a p
ortion of that field as a final resting-place for those who here
```

[3] Lincoln's speech was intended to be a dedication of the battlefield's cemetary and followed orator Edward Everett's two-hour speech. No one remembers what Everett said that November day

```
gave their lives that that nation might live. It is altogether f
itting and proper that we should do this. But in a larger sense,
we cannot dedicate, we cannot consecrate, we cannot hallow this
ground. The brave men, living and dead who struggled here have c
onsecrated it far above our poor power to add or detract. The wo
rld will little note nor long remember what we say here, but it
can never forget what they did here. It is for us the living rat
her to be dedicated here to the unfinished work which they who f
ought here have thus far so nobly advanced. It is rather for us
to be here dedicated to the great task remaining before us--that
from these honored dead we take increased devotion to that cause
for which they gave the last full measure of devotion--that we h
ere highly resolve that these dead shall not have died in vain,
that this nation under God shall have a new birth of freedom, an
d that government of the people, by the people, for the people s
hall not perish from the earth."
```

As we have discussed, we need a way to represent text data for machine learning algorithm and although we can easily analyze Lincoln's dedication address, it will be instructive to do this analysis using Python code. The technique that follows is referred to as the **bag-of-words** model helps us to achieve that task (we'll have more to say about this in Chapter 4).

Definition 3.5. *The bag-of-words model is a simplifying representation used in NLP to represent text (such as a words, sentences, or documents) as the bag (multiset) of its words, disregarding grammar and even word order, but keeping multiplicity.*

The bag-of-words model is simple to understand and implement. It is a way of extracting features from the text for use in machine learning algorithms. In our approach, we use sentences from the speech as the tokens (there are ten).

```
dataset = nltk.sent_tokenize(text)
for i in range(len(dataset)):
    dataset[i] = dataset[i].lower()
    dataset[i] = re.sub(r'\W', ' ', dataset[i])
    dataset[i] = re.sub(r'\s+', ' ', dataset[i])
dataset
```

```
[1]'fourscore and seven years ago our fathers brought forth on t
his continent a new nation conceived in liberty and dedicated to
the proposition that all men are created equal ',
```

```
[2] 'now we are engaged in a great civil war testing whether tha
t nation or any nation so conceived and so dedicated can long en
dure ',
[3] 'we are met on a great battlefield of that war ',
[4] 'we have come to dedicate a portion of that field as a final
resting place for those who here gave their lives that that nati
on might live ',
[5] 'it is altogether fitting and proper that we should do this
',
[6] 'but in a larger sense we cannot dedicate we cannot consecra
te we cannot hallow this ground ',
[7] 'the brave men living and dead who struggled here have conse
crated it far above our poor power to add or detract ',
[8] 'the world will little note nor long remember what we say he
re but it can never forget what they did here ',
[9] 'it is for us the living rather to be dedicated here to the
unfinished work which they who fought here have thus far so nobl
y advanced ',
[10] 'it is rather for us to be here dedicated to the great task
remaining before us that from these honored dead we take increas
ed devotion to that cause for which they gave the last full meas
ure of devotion that we here highly resolve that these dead shal
l not have died in vain that this nation under god shall have a
new birth of freedom and that government of the people by the pe
ople for the people shall not perish from the earth ']
```

We treated each sentence as a separate document and we now make a list of all words from all the ten documents, while excluding the punctuation.

The next step is to get the words as tkens from each document and count their occurrence. For example, we take the first document, "fourscore and seven years ago our fathers brought forth on this continent a new nation conceived in liberty and dedicated to the proposition that all men are created equal," and we treat words of the sentence as tokens, counting them as we go to obtain their frequencies.

```
# Creating the Bag of Words model
word2count = {}
for data in dataset:
    words = nltk.word_tokenize(data)
    for word in words:
        if word not in word2count.keys():
            word2count[word] = 1
        else:
```

```
            word2count[word] += 1
word2count
```

Now, we will extract the 35 most frequently occurring words (35 is arbitray and we could extract 20, 40, or 100), using *heapq*.

```
import heapq
freq_words = heapq.nlargest(35, word2count, key=word2count.get)
freq_words
```

```
{'fourscore': 1,
 'and': 6,
 'seven': 1,
 'years': 1,
 'ago': 1,
 'our': 2,
 'fathers': 1,
.
.
.
 'birth': 1,
 'freedom': 1,
 'government': 1,
 'people': 3,
 'by': 1,
 'perish': 1,
 'earth': 1}
```

If we extract the 12 most frequent words, nearly all of them are words common to all documents.

```
import heapq
freq_words = heapq.nlargest(10, word2count, key=word2count.get)
freq_words
```

```
 'that': 13,
 'the': 11,
 'we': 10,
 'to': 8,
 'here': 8,
 'a': 7,
 'and': 6,
```

```
'nation': 5
'can': 5,
'of': 5,
'have': 5,
'for': 5,
```

The words that are common to all documents, even documents not in our corpus, are called **stop words** or **stopwords**. Using Python, we can filter wout the stopwords.

```
from nltk.corpus import stopwords
key_words = []
en_stops = set(stopwords.words('english'))
for word in freq_words:
    if word not in en_stops:
        key_words.append(word)
print(key_words)
```

```
['nation', 'dedicated', 'great', 'dead', 'us', 'shall', 'people
', 'new', 'conceived', 'men', 'war', 'long']
```

Now, we vectorize the corpus containing only the key words:

```
X = []
for data in dataset:
    vector = []
    for word in key_words:
        if word in nltk.word_tokenize(data):
            vector.append(1)
        else:
            vector.append(0)
    X.append(vector)
X = np.asarray(X)
print(X)
```

```
array([[1, 1, 0, 0, 0, 0, 0, 1, 1, 1, 0, 0],
       [1, 1, 1, 0, 0, 0, 0, 0, 1, 0, 1, 1],
       [0, 0, 1, 0, 0, 0, 0, 0, 0, 0, 1, 0],
       [1, 0, 0, 0, 0, 0, 0, 0, 0, 0, 0, 0],
       [0, 0, 0, 0, 0, 0, 0, 0, 0, 0, 0, 0],
       [0, 0, 0, 0, 0, 0, 0, 0, 0, 0, 0, 0],
       [0, 0, 0, 1, 0, 0, 0, 0, 0, 1, 0, 0],
       [0, 0, 0, 0, 0, 0, 0, 0, 0, 0, 0, 1],
       [0, 1, 0, 0, 1, 0, 0, 0, 0, 0, 0, 0],
       [1, 1, 1, 1, 1, 1, 1, 1, 0, 0, 0, 0]])
```

Then, document (sentence) 1 contains ['nation', 'dedicated', 0, 0, 0, 0, 0,'new', 'conceived', and 'men', 0, 0]. So, only five of the 12 words are contained in document 1 of the filtered text. We can visualize this better with a graphic, and we use a heatmap in this instance in Figure 3-2.

```
import seaborn as sns
import matplotlib
from matplotlib import pyplot as plt
%matplotlib inline #generate plots in the notebook
fig, ax = plt.subplots(figsize=(16,8))
sns.heatmap(X, annot=True, annot_kws={"size": 12}, xticklabels=k
ey_words)
plt.xticks(fontsize=12)
plt.yticks(fontsize=12)
plt.ylabel('Sentence', fontsize = 14)
plt.xlabel('Word', fontsize = 14)
plt.show()
```

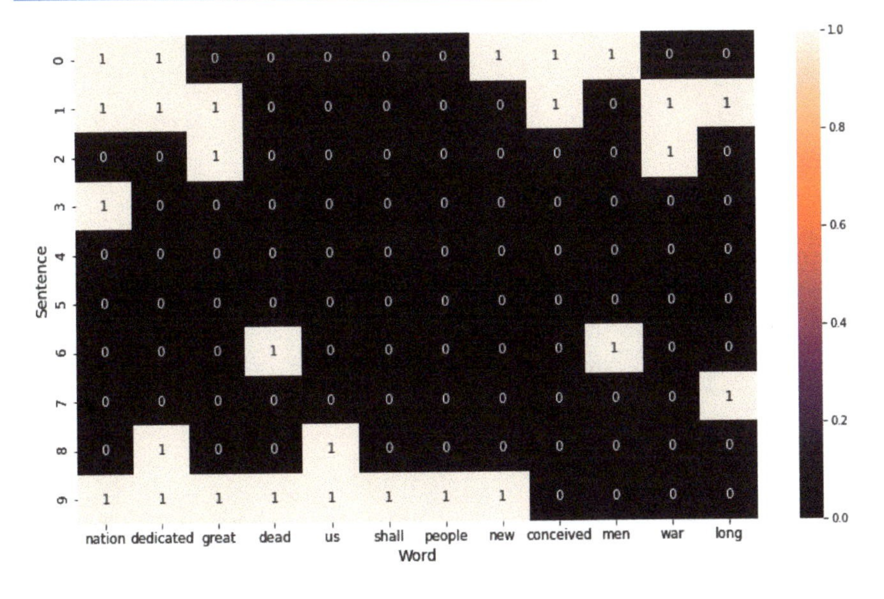

Figure 3-2. Heatmap of key words.

3.6 Summary

In summary, the process of converting NLP text into numbers is called vectorization in ML. Moreover, we saw two different ways to convert text into vectors are: (1) counting the number of times each word appears in a document, and (2) calculating the frequency that each word appears in a document out of all the words in the document.

3.7 Exercise

Take the text below and perform TF-IDF vectorization. The text is the last paragraph (closing) of Lincoln's naugural Address.

"With malice toward none with charity for all with firmness in the right as God gives us to see the right let us strive on to finish the work we are in to bind up the nation's wounds, to care for him who shall have borne the battle and for his widow and his orphan ~ to do all which may achieve and cherish a just and lasting peace among ourselves and with all nations."

Chapter 4 – Vectorizing Data Application

4.1 Introduction¶

Banks are a prime example of businesses that already do or should take advantage of data science solutions to a variety of situations, including, but not limited to:

- Call routing for complex calls
- Call routing for complaints
- Quality control for regulatory compliance
- Customer sentiment analysis
- Fraudulent calls and other contacts

If we can select words and phrase from manuscripts or by actively monitoring calls, chats, and so on, then we can use past patterns to predict future behavior. We can also gauge customer sentiment when speaking to our best member service representatives (MSR), extract the words and phrases to characterize what "good looks like," and train other MSRs using the information we've obtained. Additionally, we can use the information to flag complex callers and route them to MSRs trained to address these situations. There are many other applications, but in this chapter, we will look at banks (financial services) we might deal with complaints about particular products or services, and (1) determine the product being complained about, and (2) predict future complains based on transcripts from past complaints.

Goal: Properly route complaints based on product groups. Our goal is to produce a data science solution that routes text complaints to the correct complaint department for each product group or service that the bank offers.

The bank's business unit receives complaints in the form of free text and wants to route the complaints to one of seven different complaint departments (with a product_group name in parenthesis):

1. Bank account or service (bank_service)
2. Credit card (credit_card)
3. Credit reporting (credict_reporting)
4. Debt collection (debt_collection)
5. Lines of loans (loan)
6. Mortgage (mortgage)

We have obtained a data set with 286,362 records that contains complaint text (field name "text"), a message identifier ("complaint_id") and a verified product group complaint department ("product_group")

4.2 Methodology¶

We have obtained a data set with 286,362 records that contains complaint text (field name "text"), a message identifier ("complaint_id") and a verified correct complaint department ("product_group").

To start, we will use a variety of techniques to explore and process the data, but will essentially follow the typical natural language processing (NLP) methodology, we discussed in Chapter 1:

1. Sentence Tokenization
2. Word Tokenization
3. Text Lemmatization and Stemming
4. Stop Words
5. Regular Expression (Regex)
6. Term Frequency-Inverse document Frequency (TF-IDF)
7. Bag-of-Words (BOW)

In this analysis, we use TF-IDF for two different functions, n-gram extraction and model fitting. For corpus feature extraction, we will create five different classification models (or text classifiers) and compare them using a set of metrics. The classifiers include:

- Multinomial Naïve Bayes
- RidgeClassifier

- SGDClassifier
- Multinomial Logistic Regression
- Random Forest (RF)
- Support Vector Machine (SVM)
- Multi-layer Perceptron Classifier (neural network)

For peforming NLP, we will use the bag-of-words (BOW) method.

4.3 Load and Reading Data¶

As is the case with all of our Python usage so far, we preload packages we think we will need, as one style of doing it. This has several advantages: (1) we will be prompted to load updates so that we are using the current ones; (2) some of the package are global (used throughout), like matplotlib; and (3) it helps us think about our process as we begin. We will also load packages later as we need may them. So, we are loading several pakages, like Scikit-learn (*sklearn*) that we will describe in sections 4.4, and following.

```
#Standard packages for loading and manipulating data
import pandas as pd
import numpy as np

# Scikit Learn for vectorization and feature extraction
from sklearn import preprocessing
from sklearn import tree
from sklearn.feature_extraction.text import CountVectorizer, Tfi
dfVectorizer, TfidfTransformer
from sklearn.model_selection import train_test_split, KFold

# Natural Language Toolkit (NLTK) for NLP functions
import nltk
from nltk.corpus import stopwords
from nltk.stem.snowball import SnowballStemmer
from nltk.corpus import wordnet as wn
from nltk.tokenize import word_tokenize
from nltk.stem import WordNetLemmatizer
lemmatiser = WordNetLemmatizer()

# Mathplotlib and Seaborn for Plotting functions like histograms
import matplotlib
from matplotlib import pyplot as plt
import seaborn as sns
```

```
# Allow plots in Notebook
%matplotlib inline

# Warning supression
import warnings
warnings.filterwarnings("ignore",category=DeprecationWarning)
```

4.3.1 *Loading (reading) the Dataset using Pandas*

Our first step in the modeling process is to load the data. In this instance the data is contained in a CSV file that we will read into the Jupyter Notebook.

```
df = pd.read_csv("D:/Documents/Data/case_study_data_copy.csv")
```

4.4 Data Exploration

The next several step we take are for exploring the complaint data, including listing the headings, viewing some records, and assessing the shape of the data frame. Often neglected, the shape of the data frame (or the dimensions of the array) will become important when we vectorize the independent variable, the text of the complaints. Also, it is a good idea to check that the data was loaded properly by examining a few records of rows of the data.

```
list(df), df.shape
```

```
(['complaint_id', 'product_group', 'text'], (268380, 3))
```

```
df.head(5) # for showing a snapshot of the dataset
```

	complaint_id	product_group	text
0	2815595	bank_service	On check was debited from checking account and...
1	2217937	bank_service	opened a Bank of the the West account The acc...
2	2657456	bank_service	in nj opened a business account without autho...

3	1414106	bank_service	A hold was placed on saving account because in...
4	1999158	bank_service	Dear CFPBneed to send a major concerncomplaint...

4.4.1 *Prepare a Frequency Distribution*

Our next step is to gain an understanding of the data and its composition. We can do this by looking at the dependent variable, product groups. We can star with examining the frequency distribution of complaints by product. A good way to do this is studying the frequency distribution, or the count of complaints by group. Let's first examine the frequency distribution numerically. There are several ways to do this with Python. One way is to calculate the length of each complaint by product group, as shown below.

```
# Provides calculation of the shape of the data for each product group
bank_service_len=df[df['product_group']=='bank_service'].shape[0]
credit_card_len=df[df['product_group']=='credit_card'].shape[0]
credit_reporting_len = df[df['product_group'] == 'credit_reporting'].shape[0]
debt_collection_len = df[df['product_group'] == 'debt_collection'].shape[0]
loan_len = df[df['product_group'] == 'loan'].shape[0]
mortgage_len = df[df['product_group'] == 'mortgage'].shape[0]
```

```
# Returns that complaint frequencies for each product group
bank_service_len, credit_card_len, credit_reporting_len, debt_collection_len, loan_len, money_transfers_len, mortgage_len
```

Another way to getthe frequencies is to call them directly as we have below.

```
df_freq = df.groupby('product_group').text.count()
df_freq
```

```
product_group
bank_service        20071
credit_card         29553
credit_reporting    81234
debt_collection     61471
loan                31036
```

```
mortgage            40281
Name: text, dtype:  int64
```

4.4.2 Plotting the Data

Now, we can plot the data and study using a histogram as shown in Figure 4-1. Again, there are several different plots and a number of methods within each. Here, we will work with histograms and look at three way to implement them with Python.

```
plt.figure(figsize = [8,6])
df.groupby('product_group').text.count().plot.bar(ylim = 0)
plt.grid(axis = 'y', alpha = 0.75)
plt.xticks(fontsize = 12)
plt.yticks(fontsize = 12)
plt.ylabel('Number of samples', fontsize = 14)
plt.xlabel('Product Group', fontsize = 14)
plt.title('Frequencies of Complaints by Product Group', fontsize
= 16)
plt.show()
```

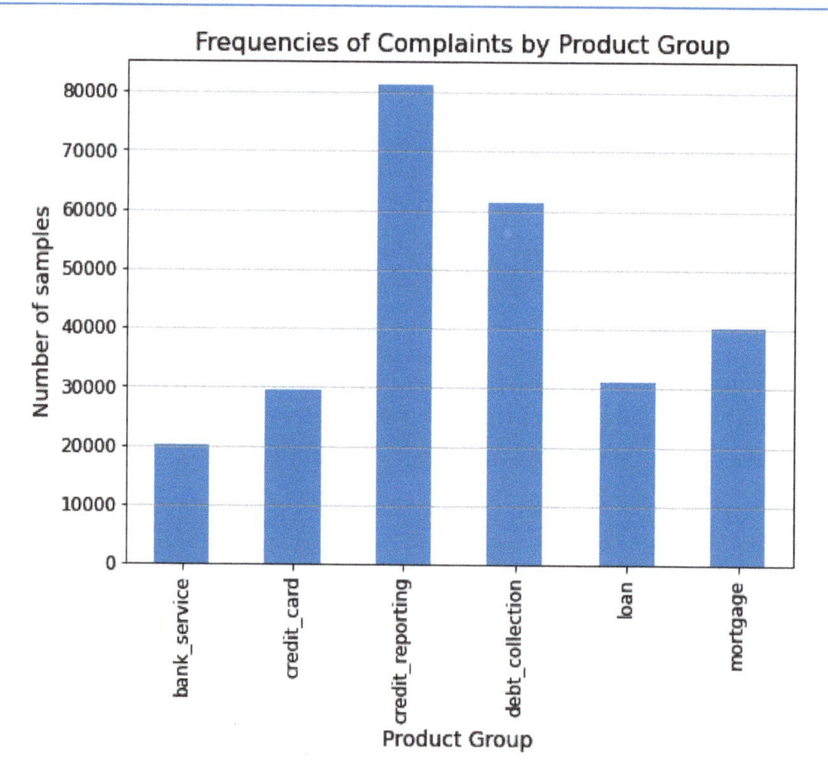

Figure 4-1. Histogram of complaints by bank product group.

4.4.3 *Histogram with Legend using Matplotlib Only*

Here we use the categories we defined above to add a legend, as well as chart title and axes labels. Using the previous method does not allow the addition of a legend. The print syntax is a little different. The first argument for `plt.bar()` is where on the x-axis we place the bar (using 1-7 for the seven product groups). This positioning is arbitrary. For instance, we can plot the bank service bar as 5 instead of 1 and the frequencies do not change; however, it is a better practice to maintain the same order for the classes and we have been using alphabetical order. The second is the component we are plotting, a product group. The third is the width of the bar. The fourth is the label for the legend. The plot is shown in Figure 4-2.

```
# Sets up the data for construct the frequency distribution (bar
chart)
plt.figure(figsize = [9,6])
plt.bar(1,bank_service_len,0.75, label = "bank_service")
plt.bar(2,credit_card_len,0.75, label = "credit_card")
plt.bar(3,credit_reporting_len,0.75, label = "credit_reporting")
plt.bar(4,debt_collection_len,0.75, label = "debt_collection")
plt.bar(5,loan_len,0.75, label = "loan")
plt.bar(6,money_transfers_len,0.75, label = "money_transfers")
plt.bar(7,mortgage_len,0.75, label = "mortgage")
plt.legend()
plt.grid(axis = 'y', alpha = 0.75)
plt.xticks(fontsize = 12)
plt.yticks(fontsize = 12)
plt.ylabel('Number of samples', fontsize = 14)
plt.xlabel('Product Group', fontsize = 14)
plt.title('Frequencies of Complaints by Product Group', fontsize
= 16)
plt.show()
```

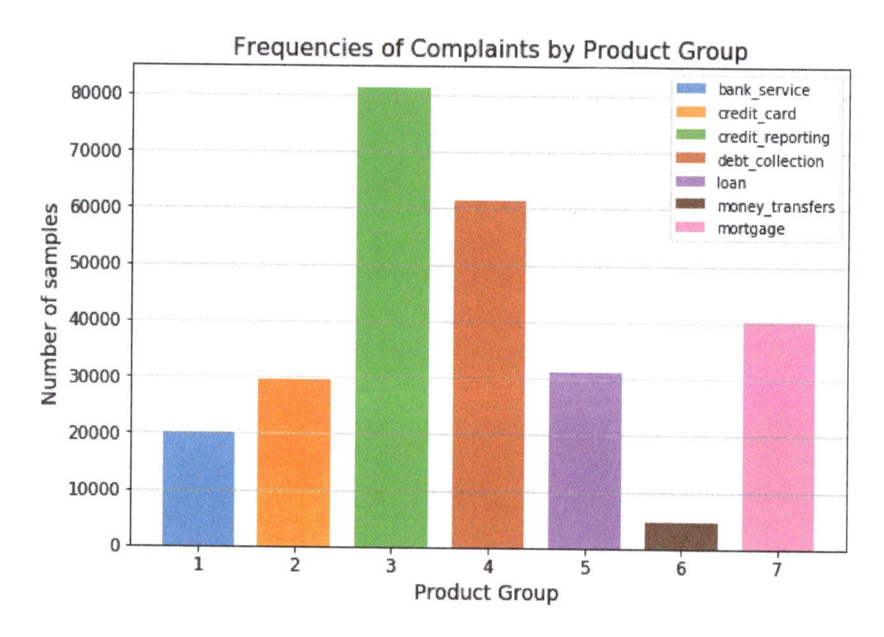

Figure 4-2. Color-coded histogram of complaints by bank product.

4.4.4 Store Product Group Data

The next step, and optional one, is to store the data for each product group into its own object for later use (potentially). In doing so, we can now call out an individual product group to analyze. for example, if we want to see the first transcript corresponding to the product group, mortgages, and the fifth complaint corresponding to the product group, credit card, we write mtg[1] followed by ccd[5].

```
bks = df[df.product_group == "bank_service"]["text"].values
mtg = df[df.product_group == "mortgage"]["text"].values
crp = df[df.product_group == "credit_reporting"]["text"].values
ccd = df[df.product_group == "credit_card"]["text"].values
lon = df[df.product_group == "loan"]["text"].values
dct = df[df.product_group == "debt_collection"]["text"].values
mts = df[df.product_group == "money_transfers"]["text"].values
```

```
mtg[1], ccd[5]
```

```
("We paid off our mortgage with  on On the evening of received a
letter stating that had to call them to cancel auto withdrawal o
f mortgage payment checked account and noticed that they had tri
ed to withdraw the mortgage payment so we were charged insuffici
ent funds charge called them and told them to cancel auto withdr
```

```
awal On they tried to withdraw again We got another charge plus
a recurring insufficient funds charge of So now we 're up to dol
lars in our account called  again and the girl kept saying have
no record of your call you never told us not to take the money a
nd there nothing we can do about it you need to put a stop payme
nt on your account Also that they would refund the fees So calle
d bank and told them what was happening and she told me that sur
e they could put stop payments on those amounts but it would cos
t for each At this point told her to close account She said she
could close it until paid the fees ",
 ' received Double Cash credit card statement for the month of w
hich included a late fee of $25 for a payment they indicated the
y received on payment was due automatic bank withdraw was made o
n and bank statement indicated the payment was received and clea
red by the on did not post payment until one day after the due d
ate and attempted to charge me the $25 late fee It took me at le
ast minutes of arguing with a supervisor who was anything but he
lpful to resolve this matter feel they have been dishonest and h
oped would not notice the fee andor not bother to complain about
it How many other people have had this experience?')
```

4.4.5 Define Variables

As we go forward with our analysis, we'll want to treat the data as having two variables, the text variable, X, is the independent variable and the product group, y, is the dependent variable. Since Python is case sensitive, we want an upper-case X and a lower-case y. We use this convention because y is a vector and X is generally an array or matrix.

```
X = corps['text']
y = corps['product_group']
```

4.4.6 Label Encoding of Classes:

This is a classification problem where the classes are the seven product groups as already mentioned. In our dataset, the class labels are non-numeric (bank_services, credit_card, credit_reporting, debt_collection, loan, money_transfers, and mortgage). To perform some quick exploratoy analysis we use LabelEncoder from Scikit-Learn to make the classes numeric, starting from 0 depicting each label in the alphabetic order i.e., (0 → bank_services, 1 → credit_card, 2 → credit_reporting, 3 → debt_collection, 4 → loan, 5 → money_transfers, and 5 → mortgage).

```
# Importing necessary libraries
from sklearn.preprocessing import LabelEncoder
labelencoder = LabelEncoder()
yL = labelencoder.fit_transform(y)
print(labelencoder.fit_transform(y))
```

```
[0 0 0 ... 5 5 5]
```

Now we created another type of histogram using the *seaborn* library of plotting functions. The result is depicted by Figure 4-3.

```
# Import the libraries
import matplotlib.pyplot as plt
import seaborn as sns

# matplotlib histogram
plt.figure(figsize = [8,6])
plt.hist(yL, color = 'lightgreen', edgecolor = 'black',bins = 6)
# seaborn histogram
sns.distplot(yL, hist = True, kde = False,
             bins = None, color = 'blue',
             hist_kws = {'edgecolor':'black'})
# Add labels
plt.grid(axis = 'y', alpha = 0.75)
plt.xticks(fontsize = 12)
plt.yticks(fontsize = 12)
plt.title('Scores',fontsize = 15)
plt.xlabel('Product Groups',fontsize = 15)
plt.ylabel('Frequencies',fontsize = 15)
```

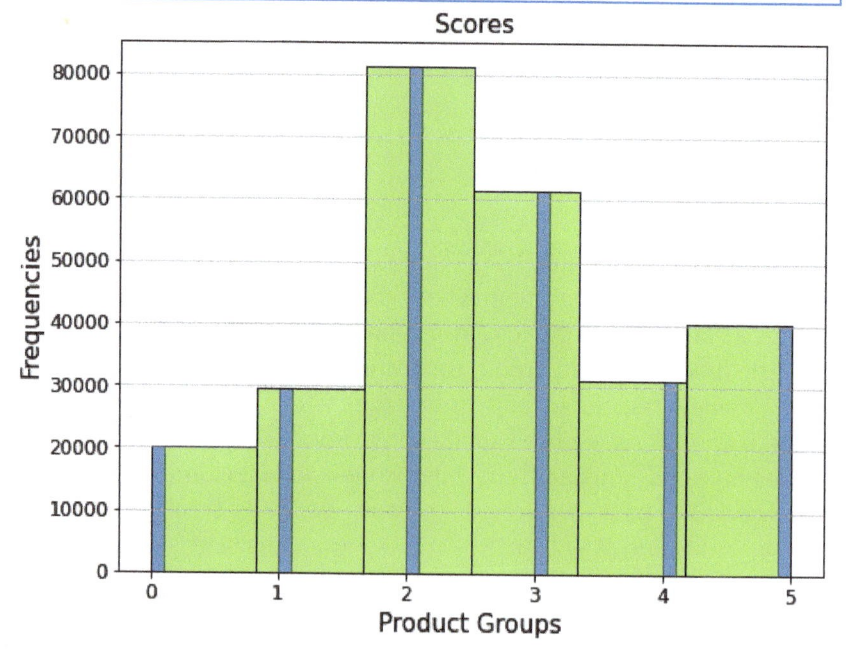

Figure 4-3. Histogram using the seaborn library of plotting functions.

4.4.7 Word Cloud Visualization:

Word clouds are a way to show the frequency of words as they occur in the dependent variable, X. We have developed the word clouds (shown in Figure 4-4 through **Figure 4-10**) as follows:

- Product groups have their own unique words and phrases, as well as some common ones
- Visualization of the mostly-used words to the least-used words for the product groups can be demonstrated
- Seven text snippets, each belonging to the seven product groups, respectively, can render a Word Cloud

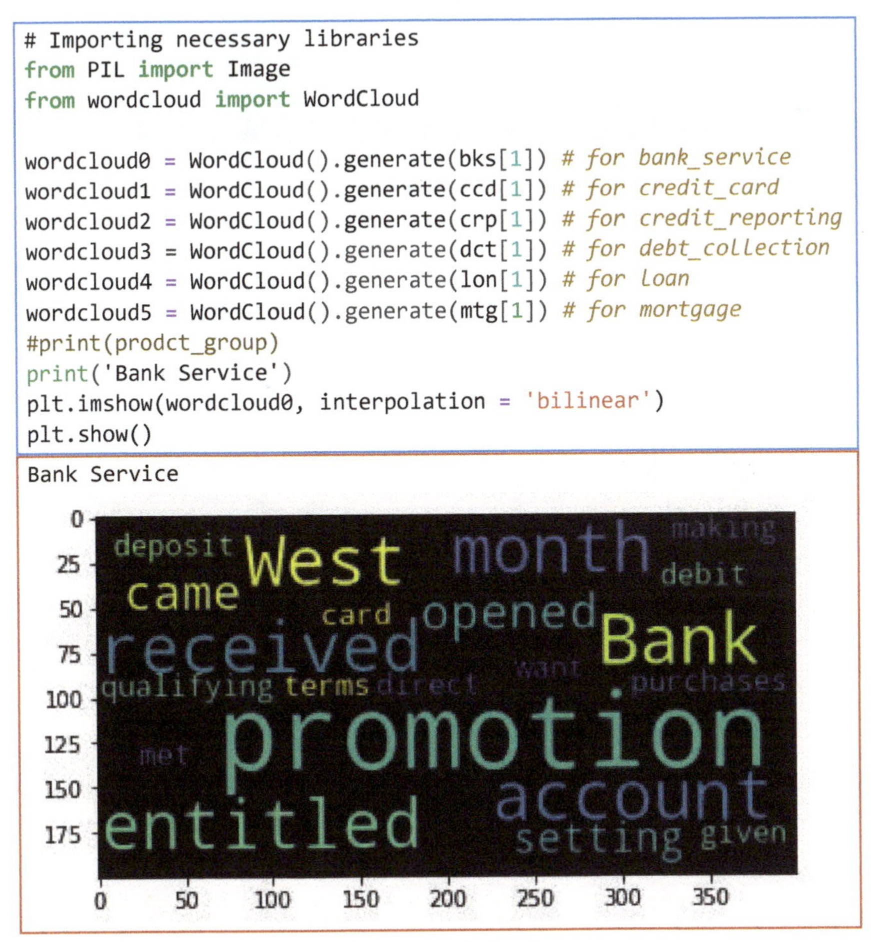

```
# Importing necessary libraries
from PIL import Image
from wordcloud import WordCloud

wordcloud0 = WordCloud().generate(bks[1]) # for bank_service
wordcloud1 = WordCloud().generate(ccd[1]) # for credit_card
wordcloud2 = WordCloud().generate(crp[1]) # for credit_reporting
wordcloud3 = WordCloud().generate(dct[1]) # for debt_collection
wordcloud4 = WordCloud().generate(lon[1]) # for loan
wordcloud5 = WordCloud().generate(mtg[1]) # for mortgage
#print(prodct_group)
print('Bank Service')
plt.imshow(wordcloud0, interpolation = 'bilinear')
plt.show()
```


Figure 4-4. Bank Service complaints word cloud.

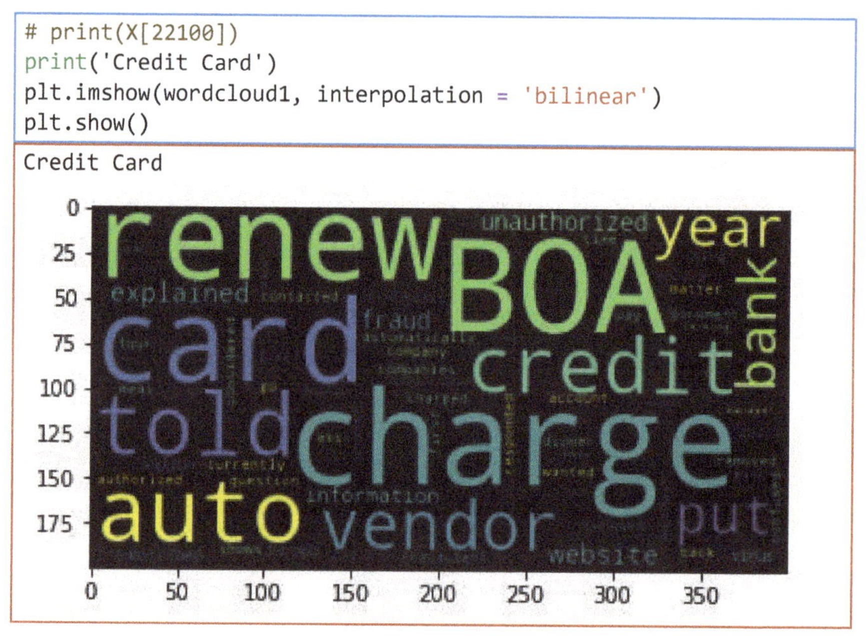

Figure 4-5. Credit Card complaints word cloud.

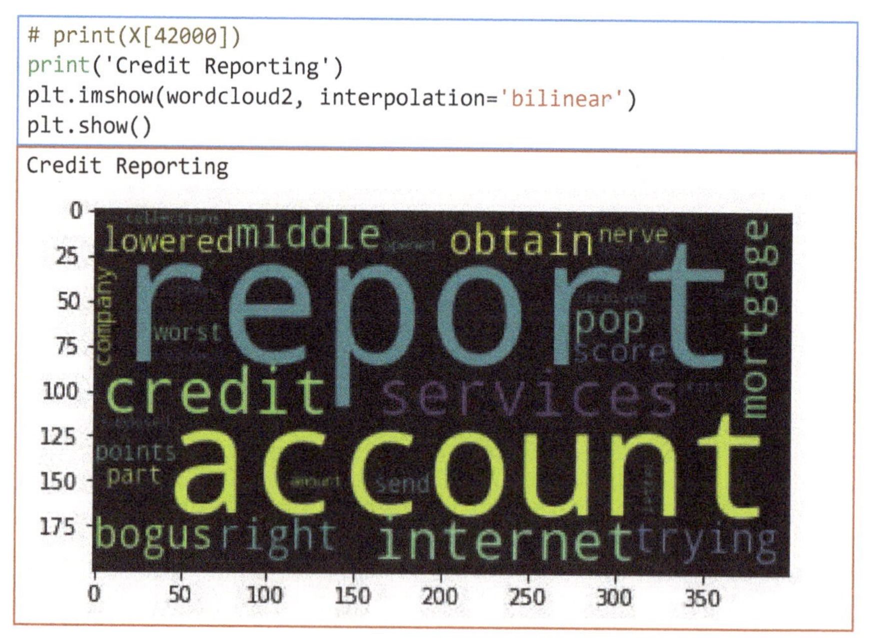

Figure 4-6. Credit Reporting complaints word cloud.

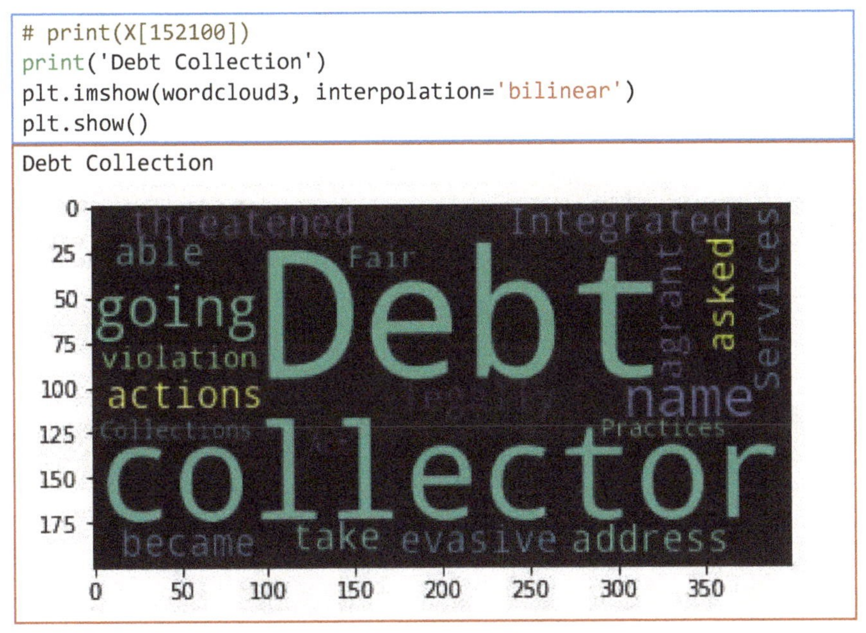

Figure 4-7. Debt Collection complaints word cloud.

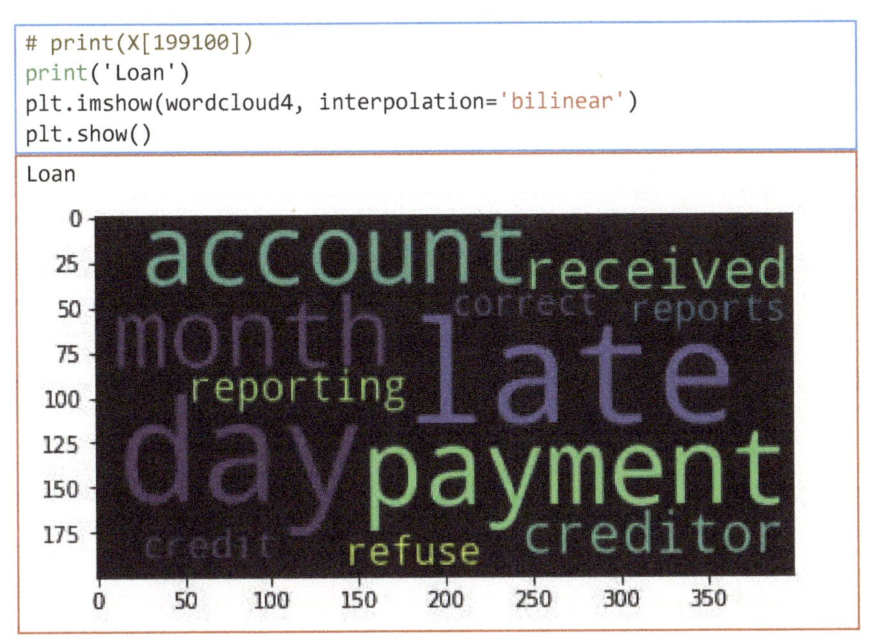

Figure 4-8. Loan complaints word cloud.

```
# print(X[227100])
print('Money Transfers')
plt.imshow(wordcloud5, interpolation='bilinear')
plt.show()
```

```
Money Transfers
```

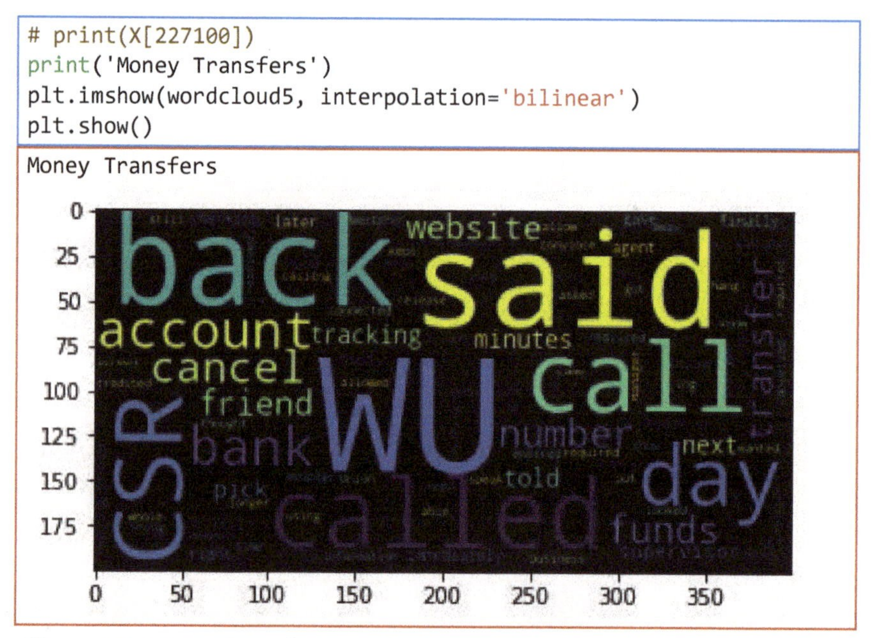

Figure 4-9. Money Transfers complaints word cloud.

```
# print(X[237100])
print('Mortgage')
plt.imshow(wordcloud6, interpolation='bilinear')
plt.show()
```

```
Mortgage
```

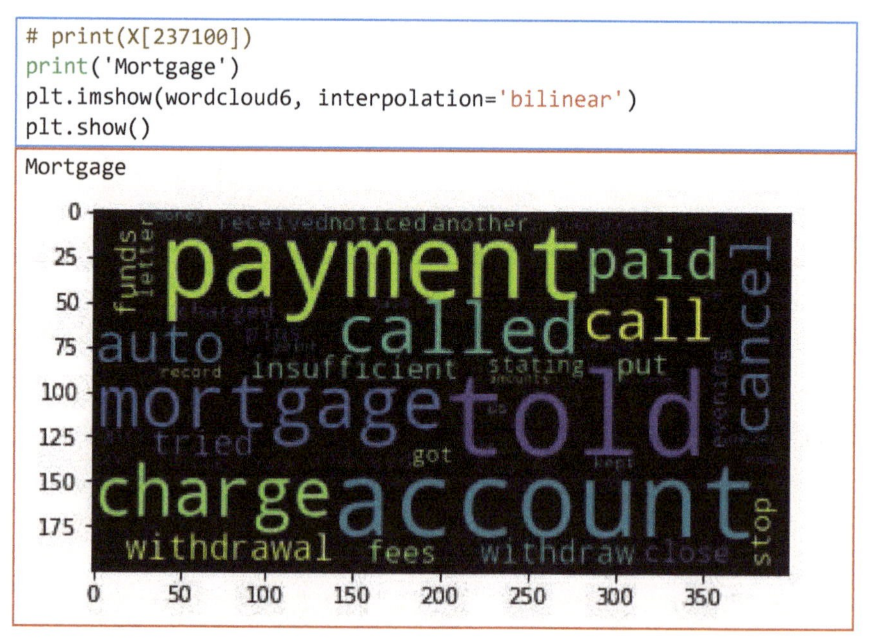

Figure 4-10. Mortgage complaints word cloud.

4.5 Natural Language Processing (NLP)

In this section we will examine how various natural language processing techniques can provide productive insights into large collections of text data. In this instance, we continue to analyze the complaint data but with much more robust measures and results. As we have already posed, NLP is the process of turning human speech (usually in written form) into language a computer can understand (numerical) using a program, extracting information from the numerical representation of the language, and turning it back into informative human language.

We looked at the basic text preparation steps in Chapter 1. We add four more step for feature extraction here:

1. Remove punctuation
2. Remove special characters
3. Remove stop words
4. Tokenize the text
5. Lemmatize the text
6. Vectorize the text
7. Perform text transformations
8. Extract n-grams

Here, will make better use of the power of Python and supporting packages, like Scikit-Learn (sklearn). These 'utilities" allow us to take a given text or collection of texts, a corpus), remove punctuation, remove whitespaces, convert to lower case, remove common words that yield no information (stopwords) to produce a "clean" version of the text we want to analyze. we will begin by discussing stopwords.

4.5.1 Stop Words

A stop word or stopword is a commonly used word (such as “the”, “a”, “an”, “in”) that a search engine has been programmed to ignore, both when indexing entries for searching and when retrieving them as the result of a search query.

There are several different ways to process text for stopwords with Python, including the Natural Language Tool Kit (*nltk*) and Skikit-Learn (*sklearn*). We could also build our own list and implement it by wring Python code to remove stopwords using our list, of adapting a list and adding or taking from it for our use. However, processing text for stopwords is not simply a matter of pressing the Run button.

The are many existing stopwords lists and in many languages. In at least one of these lists is the word "computer," a word you may not want to remove, when analyzing text about competing technologies. Likewise, with save when analyzing retail banking issues. Stopwords from Skikit-Learn can be called after importing them, while NTLK stopwords are called within the `CountVeterizor()` function, which we will get to later in this section.

When we invoke stopwords with `CountVectorizer`, one of the arguments is, stopwords = None, as the default. If we want English stopwords, we merely add the argument, stopwords = 'english'.

4.5.2 Customized Stop Words

We will defer implementing the removal of stopwords until our discussion of `CountVectorizer`. For now, let's look at a list of stopwords I adapted from R. I added 'non-words' like "uh-huh", "um", "lol", and do on. In some settings, these might be appropriate words for analysis, but we usually exclude them and others like them.

```
stopwords = pd.read_csv("D:/Documents/Data/english_stopwords.txt
")
print(stopwords)
```

```
Empty DataFrame
Columns: ['i', 'll', 'me', 'my', 'myself', 'we', 'our', '
ours', 'ourselves', 'you', "you're", "you've", "you'll", "
you'd", 'your', 'yours', 'yourself', 'yourselves', 'he', '
him', 'his', 'himself', 'she', "she's", 'her', 'hers', 'h
erself', 'it', "it's", 'its', 'itself', 'they', 'them', '
their', 'theirs', 'themselves', 'what', 'which', 'who', 'w
hom', 'this', 'that', "that'll", 'these', 'those', 'am',
'is', 'are', 'was', 'were', 'be', 'been', 'being', 'have'
, 'has', 'had', 'having', 'do', 'does', 'did', 'doing',
'a', 'an', 'the', 'and', 'but', 'if', 'or', 'because', '
as', 'until', 'while', 'of', 'at', 'by', 'for', 'with',
'about', 'against', 'between', 'into', 'through', 'during',
'before', 'after', 'above', 'below', 'to', 'from', 'up',
'down', 'in', 'out', 'on', 'off', 'over', 'under', 'again
', 'further', 'then', ...]
Index: []
[0 rows x 183 columns]
```

```
# Importing necessary libraries
# Natural Language Toolkit
import nltk
from nltk.corpus import stopwords
```

4.5.3 Lemmatization

Inflected forms of a word are known as lemma. For example, "studying," "studied" are inflected forms or lemma of the word study which is the root word. So, the lemma of a word is grouped under the single root word. This is done to make the vocabulary of words in the corpus contain distinct words only.

Lemmatization is the process of converting a word to its base form. Lemmatization igoes beyond stemming, since lemmatization considers the context and converts the word to its meaningful base form--stemming just removes the last few characters, often leading to incorrect meanings and spelling errors.

4.5.4 Create Train and Test Sets

Before we proceed further, we will setup training and testing sets that we will use for feature selection with the classification methods we pre-selected (Multinomial Naive Bayes, Support Vector Machines, etc.). Once we have these sets defined, with will begin working on feature extraction.

We will use the Scikit-Learn modules model.selection.train_test_split to establish these sets. The module splits arrays or matrices into random train and test subsets, by setting the size (percentage) of the test set. Here we want a training set that is comprised by 70&% of the test data, with the remaining 30% held as the testing set. Also, we want to have separate sets for the text, X (complaints), and the categories, y (product groups), as we will only need to prepare the text, X, using NLP. We will separately prepare the dependent variable, y, converting the word-classes to numeric categories.

Apart from the import statement, the first for steps have been performed previously. We repeat them here to to refresh the data for our NLP work. This fifth line is the code for splitting X and y to for the train and ets sets.

```
from sklearn.model_selection import train_test_split
dta = pd.read_csv("D:/Documents/Data/case_study_data_copy.csv")
corps = dta[['product_group','text']]
```

```
X = corps.text
y = corps.product_group
X_train, X_test, y_train, y_test = train_test_split(X, y, test_s
ize=0.3, random_state=1234)
```

4.5.5 Feature Extraction

The Scikit-Learn library, *sklearn.feature_extraction*, can be used to extract features in a format supported by machine learning algorithms from datasets consisting of formats such as text and image. Formally, **feature extraction** is a process of dimensionality reduction by which an initial set of raw data is reduced to more manageable groups for processing. However, this is not the same thing as **feature selection**, which is our goal when modeling, particularly when we want to predict future outcomes based on past data and feature selection. In our analysis here, we are just examining the text to see features exist that could help classify the text (complaints) that are unique to each (or any) of the product groups.

> ***Definition 4.1.*** *Feature Selection is the process where you automatically or manually select those features which contribute most to your prediction variable or output in which we are interested.*

4.5.6 Vectorization

Vectorization is the process of converting data into a mathematical construct, like a vector, that can be easily manipulated without changing its content. Vectorization also speeds up computation. We usually evaluate the efficiency of a program by how fast it can manipulate large data sets in a relatively short time interval. Python uses mathematical functions for fast operations on entire arrays of data without having to write loops, and consequently produce results faster.

4.5.7 CountVectorizer

CountVectorizer converts text to word count vectors. It provides a simple way to both tokenize a collection of text documents, like our collection of complaints, and build a vocabulary of frequently used words, where the frequency can range from 1… n, known words. This vocabulary of known words can then be used to encode new related documents for analysis. In this manner, we can use a collection of text (complaints) to predict complaints based on this encoding. In our particular case, we will encode the text in such a manner as to not only predict complaints, but to categorize those complaints as belonging to one of the banks services/products, i.e., loans, credit cards, debt collection, and so on. Here, we will use *CountVectorizer* to:

1. Create an instance of the **`CountVectorizer()`** function
2. Call the **`fit()`** function to learn a vocabulary from one or more documents
3. Call the **`transform()`** function on one or more documents as needed to encode each as a vector

The pseudocode for this might appear as:

- Create the transform using vectorizer = ***`CountVectorizer()`***
- Tokenize and build the vocabulary with ***`vectorizer.fit(text)`***
- Summarized the results of vectorization using print(***`vectorizer.vocabulary_`***) [this function accesses the vocabulary to see what exactly was tokenized]
- Encode the document with vector = ***`vectorizer.transform(text)`***
- Summarize the encoded vector using **`print(vector.shape)`**, ***`print(type(vector))`***, and ***`print(vector.toarray())`***

An encoded vector is returned with a length of the entire vocabulary and an integer count for the number of times each word appeared in the document, hence the "count" in vectorize. Now, these vectors will contain a lot of zeros, which we call them sparse vectors. However, Python provides an efficient way of handling sparse vectors using the *scipy.sparse* package. *SciPy* builds on this the *numpy* array object we dicussed in section 2.4.3, and is part of the *NumPy* stack which includes tools like *Matplotlib* and *pandas*. Scipy provides a large number of functions that operate on *numpy* arrays and are useful for different types of applications.

The vectors returned from a call to **transform()** function are sparse vectors, and we can transform them back to *numpy* arrays. We can then examine them to better understand what occurs by using the **toarray()** function.

```
from sklearn.feature_extraction.text import CountVectorizer

CountVectorizer(
# the input text data or corpus
    input='corps',
# encoding is used to decode
    encoding='utf-8',
# means a UnicodeDecodeError will be raised (other values are ig
nored and replace)
    decode_error='strict',
# removes accents and perform other character normalization (asc
ii is the fastest)
    strip_accents='ascii',
# converts all text to lower case
    lowercase=True,
# default value is None (only applies if analyzer == 'word')
    tokenizer=word_tokenize,
# default value is None (only applies if analyzer == 'word')
    stop_words='english',
# string, denoting what constitutes a "token" (only used if anal
yzer == 'word')
    token_pattern=r'\b\w+\b',
# will yield unigrams, bigrams, and trigrams
    ngram_range=(1, 3),
# feature makeup {'string','word','char','char_wb'} or callable
    analyzer='word',
# ignore terms that have a frequency higher than this threshold
    max_df=1.0,
# ignore terms that have a frequency lower than this threshold
    min_df=1,
# build a vocabulary size N or None
    max_features = 20,
# if True, all non zero counts are set to 1
    binary=False,
# type of the matrix returned by fit_transform() or transform()
    dtype= np.int64
)
CountVectorizer(analyzer='word', binary=False,
decode_error='strict',
dtype=<class 'numpy.int64'>, encoding='utf-8',
          input='corps',
         lowercase=True, max_df=1.0, max_features=20, min_df=1,
         ngram_range=(1, 3), preprocessor=None,
          stop_words='english',
         strip_accents='ascii', token_pattern='\\b\\w+\\b',
         tokenizer=<function word_tokenize at 0x000001A37170C0D0>
          ,vocabulary=None)
```

4.5.8 Term Frequency–Inverse Document Frequency

The `TfidfTransformer` function transforms a count matrix to a normalized term frequency (tf) or term frequency times inverse document frequency (tf-idf) representation. This is a common term weighting scheme in information retrieval, that has also found good use in document classification. The goal of using tf-idf instead of the raw frequencies of occurrence of a token in a given document is to scale down the impact of tokens that occur very frequently (see **Definition 3.4**). We use this transform because frequently occurring tokens in a given corpus are empirically less informative than features that occur in a small fraction of the training corpus.

The formula that is used to compute the tf-idf for a term *t* of a document *d* in a document set is *tf-idf(t, d) = tf(t, d) idf(t), and the* idf *is computed as idf(t) = log [n / df(t)] + 1 (if smooth_idf=False),* where *n* is the total number of documents in the document set and *df(t)* is the document frequency of *t*; the document frequency is the number of documents in the document set that contain the term *t*. The effect of adding "1" to the idf in the equation above is that terms with zero *idf*, i.e., terms that occur in all documents in a training set, will not be entirely ignored. (Note that the *idf* formula above differs from the standard textbook notation that defines the *idf as*

$$idf(t) = \log\,[\frac{n}{df(t)+1}]) *$$

```
from sklearn.feature_extraction.text import TfidfTransformer
TfidfTransformer(
# each output row will have unit norm, either:'l2','l1', or None
    norm ='l2',
    use_idf=True, #enable inverse-document-frequency reweighting
    smooth_idf=True, #smooth idf weights to prevent divison by 0
    sublinear_tf=False, #apply sublinear tf scaling
)
TfidfTransformer(norm='l2', smooth_idf=True, sublinear_tf=False,
use_idf=True)
```

4.6 Discrete Classifiers

In this section, we will define the primary building blocks of the metrics we'll use to evaluate classification models. But first, a fable:

An Aesop's Fable: The Boy Who Cried Wolf (compressed)

A shepherd boy gets bored tending the town's flock. To have some fun, he cries out, "Wolf!" even though no wolf is in sight. The villagers run to protect the flock, but then get really mad when they realize the boy was playing a joke on them. One night, the shepherd boy sees a real wolf approaching the flock and calls out, "Wolf!" The villagers refuse to be fooled again and stay in their houses. The hungry wolf turns the flock into lamb chops. The town goes hungry. Panic ensues.

Let's make the following definitions:

- "Wolf" is a positive class.
- "No wolf" is a negative class.

We can summarize our "wolf-prediction" model using a 2x2 confusion matrix that depicts all four possible outcomes:

True Positive (TP): • Reality: A wolf threatened. • Shepherd said: "Wolf." • Outcome: Shepherd is a hero.	**False Positive (FP):** • Reality: No wolf threatened. • Shepherd said: "Wolf." • Outcome: Villagers are angry at shepherd for waking them up.
False Negative (FN): • Reality: A wolf threatened. • Shepherd said: "No wolf." • Outcome: The wolf ate all the sheep.	**True Negative (TN):** • Reality: No wolf threatened. • Shepherd said: "No wolf." • Outcome: Everyone is fine.

This leads to the following definition.

Definition 4.2. *A discrete classifier, is an algorithm which given a sample from the test data, outputs* ***yes*** *(Y or 1) or* ***no*** *(N or 0). When the classifier expects the sample to belong in supset of positives, p, it outputs Y, otherwise it outputs N.*

$$E(s) = \begin{cases} 1 & E(s) \in positives \\ 0 & E(s) \in negatives \end{cases}$$

4.6.1 Naïve Bayes Classification

Naïve Bayes is a very simple and reliable way to analyze how well text can be used to predict outcomes. In our case, we want to determine if the complaint data can be used to accurately classify and the predict the complaints by back services/product. In other words, can we use the present encoded complaint data to determine if a new complaint is correctly classified as a debt collect complaint, for example.

Here, we invoke yet another Scikit-Learn function, *MultinomialNB()*. The multinomial distribution normally requires integer feature counts. However, in practice, fractional counts such as tf-idf may also work, and we assume that here as well. *MultinomialNB()* has several functions:

- Fit function takes on two arguments: X, y or X-train, y_train when training a model.
- Get_params retrieves the parameters for this estimator.
- Predict function perform classification on an array of test vectors X, i.e., X_test.
- Predict_proba returns the probability estimates for the test vector X.
- Predict_log_proba returns the log-probability estimates for the test vector X
- Score function returns the mean accuracy on the given test data and labels.

```
from sklearn.naive_bayes import MultinomialNB
mnb_clf = MultinomialNB(
    alpha=1.0,  # additive (Laplace/Lidstone) smoothing
         parameter (0 for no smoothing).
    class_prior = None, # prior probabilities of the classes
    fit_prior = True  # whether to learn class prior
         probabilities or not, false = uniform prior used
)
```

4.6.2 Pipeline Definition

Pipeline can be used to chain multiple estimators into one. This is useful as there is often a fixed sequence of steps in processing the data, for example feature selection, normalization and classification.

The purpose of the pipeline is to assemble several steps that can be cross-validated together while setting different parameters. For this, it enables setting parameters of the various steps using their names and the parameter name separated by a '__', as in the example below. A step's estimator may be replaced entirely by setting the parameter with its name to another estimator, or a transformer removed by setting it to 'passthrough' or None.

A Pipeline provides:

- Convenience and encapsulation: allow us to call fit and predict once on our data to fit a whole sequence of estimators.
- Joint parameter selection: allow us to grid search over parameters of all estimators in the pipeline at once.
- Safety: pipelines help avoid "leaking" statistics from our test data into the trained model in cross-validation, by ensuring that the same samples are used to train the transformers and predictors.

All estimators in a pipeline, except the last one, must be transformers (i.e. must have a transform method). The last estimator may be any type (transformer, classifier, etc.).

```
from sklearn.pipeline import Pipeline
text_clf = Pipeline([
     ('vect', CountVectorizer()),
     ('tfidf_trans', TfidfVectorizer()),
     ('mnb_clf', MultinomialNB()),
 ])
```

For our problem, we have already defined `CountVectorizer()`, `TfidfTransformer()`, and `MultinomialNB()`, so we could point to those functions with their represenations, e.g., `tfidf.fit_transform(X_train)`. However, we will not need these until later. For now, we turn to an alternative approach.

4.6.3 Product-to-Category and Category-to-Product Definition

If we were just concerned with generating information about complaints, we would not have to perform this step. We would merely call the n-grams from text with the `CountVectorizer` function. However, we want to encode the text in such a manner that we can identify complaints related to one of the bank services/products. Consequentially, we need to take the complaint about debt collection, and not only tokenize and transform the complaint transcipts, but also assign a number to the category, debt collection. For example, the code we use below assigns 'zero' to the category 'bank services'. By calling the function `category_to_id.items()`, we can see all the category numeric assignments as coded below and partially shown in Table 2-1.

```
from io import StringIO
df = df['product_group', 'text']
df = df[pd.notnull(df['text'])]
df.columns = ['product_group','text']
df['category_id'] = df['product_group'].factorize()[0]
category_id_df = df[['product_group', 'category_id']].drop_dupli
cates().sort_values('category_id')
category_to_id = dict(category_id_df.values)
id_to_category = dict(category_id_df[['category_id',
'product_group']].values)
df.head()
```

Table 4-1. Bank Services is assigned category ID = 0.

	product_group	text	category_id
0	bank_service	On check was debited from checking account and...	0
1	bank_service	opened a Bank of the the West account The acc...	0
2	bank_service	in nj opened a business account without autho...	0
3	bank_service	A hold was placed on saving account because in...	0
4	bank_service	Dear CFPBneed to send a major concerncomplaint...	0

Again, by calling the function `category_to_id.items()`, we can see all the category numeric assignments:

```
category_to_id.items()
dict_items([('bank_service', 0), ('credit_card', 1), ('credit_re
porting', 2), ('debt_collection', 3), ('loan', 4), ('money_trans
fers', 5), ('mortgage', 6)])
```

4.6.4 TfidfVectorizer

This feature extraction function converts a collection of raw documents to a matrix of TF-IDF features. It is equivalent to invoking `CountVectorizer` followed by `TfidfTransformer`. The parameters for the function are the combined parameters from `CountVectorizer` and `TfidfTransformer` with the same definitions. For this problem, it makes more sense to use the `TfidfVectorizer` function. The reason for this is the process of extracting n-grams from the text requires the greater functionality provided, as we will want to use the same construct to define "features" and to pull "feature names" from. (See Extracting N-grams, below.)

```
from sklearn.feature_extraction.text import TfidfVectorizer
TfidfVectorizer(
    # the input text data or corpus
    input='corps',
    # encoding is used to decode
    encoding='utf-8',
    # means a UnicodeDecodeError will be raised (other values
    are ignored and replaced)
    decode_error='strict',
    # removes accents and perform other character normalization
    (ascii is the fastest)
    strip_accents='ascii',
    # converts all tect to lower case
    lowercase=True,
    # default value is None (only applies if analyzer == 'word')
    tokenizer=word_tokenize,
    # default value is None (only applies if analyzer == 'word')
    stop_words='english',
    # string, denoting what constitutes a "token" (only used if
    analyzer == 'word')
    token_pattern=r'\b\w+\b',
    # yields unigrams, bigrams, & trigrams
    ngram_range=(1, 3),
    # feature makeup {'string','word','char', 'char_wb'} or
    callable
    analyzer='word',
    #ignore terms that have frequency higher than this threshold
    max_df=1.0,
    # ignore terms that have frequency lower than this threshold
    min_df=1,
    # build a vocabulary size N or None
    max_features = 20,
    # if True, nonzero counts are set to 1
    binary=False,
    # type of the matrix returned by fit_transform() or
    transform()
    dtype= np.int64,
    # each output row will have unit norm: 'l2', 'l1', or None
    norm='l2',
    # enables inverse-document-frequency re-weighting
    use_idf=True,
    # smooth idf weights by adding one to document freqs to
    prevent divison by zero
    smooth_idf=True,
    #apply sublinear tf scaling, i.e., replace tf with 1+log(tf)
    sublinear_tf=False,
)
```

For the problem at hand, we will define **tfidf** as **TfidfVectorizer** with paramters as shown below, and use it to extract features, labels, and n-grams, among other pruproses, including bag-of-words (BOW). Specifically, we will use it to fit the complaints from **df.text**, to prepare the independent variable for BOW.

```
from sklearn.feature_extraction.text import TfidfVectorizer
tfidf= TfidfVectorizer(sublinear_tf=True,
                        min_df=5,
                        norm='l2',
                        encoding='latin-1',
                        ngram_range=(1, 3),
                        stop_words='english',
                        lowercase=True,
                        token_pattern=r'\b\w+\b',
                        analyzer='word',
                       )
features = tfidf.fit_transform(df.text)
labels = df.category_id
```

4.6.5 Extracting N-Grams

Extracting n-grams is made easy using **CountVectorizer()** and **TfidfVectorizer()**.

Definition 4.3. *An n-gram is a contiguous sequence of n items from a given sample of text or speech. The items can be phonemes, syllables, letters, words or base pairs according to the application.*

This problem requires a little more beyound merely calling n-grams, or extracting features. For this problem, we want to get n-grams for each complaint category, i.e., bank servic, debt collection, and so on. We have already set ourselves up for success in the previous section on Product-to-Category and Category-to-Product definition, as well as the vectorization we have already performed. Where one may run into difficulties implementing this technique is to not pay attention to dimensions or shapes after vectorization, as the vectors are going to be operated on using matrix algebra, and dimension must have the right shape for the resulting vector. The error would be most likely to occur in the first statement after "for product group..."

We have already defined **TfidfVectorizer()** and used it to fit our set of complaints (repeated below). Then, in the "for" loop, we use it with the **get_feature_name** method in a numpy array, and we get the features that are evaluated to be n-grams from the complaint transcripts. The rest of the code is for printing the output, n-grams for each product group. The n-grams we are interested in are n = 1, 2, and 3, or unigrams, bigrams, and trigrams, respectively.

```
# N-GRAMS #
tfidf= TfidfVectorizer(sublinear_tf=True, min_df = 5, norm='l2',
encoding='latin-1', ngram_range=(1, 3), stop_words='english')
features = tfidf.fit_transform(df.text) # TRANSFORM
labels = df.product_group # LABEL
from sklearn.feature_selection import chi2
import numpy as np
N = 3
for product_group, category_id in sorted(id_to_category.items())
:
  features_chi2 = chi2(features,labels == category_id)
  indices = np.argsort(features_chi2[0])
  feature_names = np.array(tfidf.get_feature_names())[indices]
  unigrams = [v for v in feature_names if len(v.split('')) == 1]
  bigrams = [v for v in feature_names if len(v.split('')) == 2]
  trigrams = [v for v in feature_names if len(v.split('')) == 3]
  print("# '{}':".format(category_id))
  print(".   Most correlated unigrams:\n.   {}".format('\n. '
.join(unigrams[-N:])))
  print(".   Most correlated bigrams:\n.   {}".format('\n. '
.join(bigrams[-N:])))
  print(".   Most correlated trigrams:\n.   {}".format('\n. '
.join(trigrams[-N:])))
```

```
# 'bank_service':
.   Most correlated unigrams:
.     deposit
.     overdraft
.   Most correlated bigrams:
.     overdraft fees
.     checking account
.   Most correlated trigrams:
.     charged overdraft fees
.     opened checking account
# 'credit_card':
.   Most correlated unigrams:
.     express
.     card
.   Most correlated bigrams:
.     american express
.     credit card
.   Most correlated trigrams:
```

```
.     credit card account
.     credit card company
# 'credit_reporting':
.   Most correlated unigrams:
.     experian
.     equifax
.   Most correlated bigrams:
.     credit file
.     credit report
.   Most correlated trigrams:
.     mistakes appear report
.     appear report understanding
# 'debt_collection':
.   Most correlated unigrams:
.     collection
.     debt
.   Most correlated bigrams:
.     collect debt
.     collection agency
.   Most correlated trigrams:
.     attempting collect debt
.     trying collect debt
# 'loan':
.   Most correlated unigrams:
.     loans
.     navient
.   Most correlated bigrams:
.     student loans
.     student loan
.   Most correlated trigrams:
.     based repayment plan
.     income based repayment
# 'mortgage':
.   Most correlated unigrams:
.     modification
.     mortgage
.   Most correlated bigrams:
.     mortgage company
.     loan modification
.   Most correlated trigrams:
.     loan servicing llc
.     ocwen loan servicing
```

The list generated n-grams that are rather intuitive, which is a desired result. This does not complete our analysis, but tells us that there are some distinguishing features for each product group complaint, which we can now further investigate.

4.7 Bag-of-Words (BOW) Preparation

4.7.1 About Bag of Words

As we discussed in Chapter 3, bag-of-words (BOW) is basically the procedure we have been been performing:

1. Clean the text
2. Tokenize the text
3. Build a vocabulary from the text
4. Generate vextors from the text

With Bag-of-Words a vocabulary of words present in the corpus is maintained, and these words serve as features for each instance or document (each complaint). For words in the current document, we use their frequencies to develop features with NLP. In this way word features are engineered or extracted from the textual data or corpus.

4.7.2 Feature Engineering using Bag-of-Words

To develop feature from text is a very relevant issue for data science. Machine learning algorithms work only on numeric data, but for the problem we are solving, data is present in the form of text only. to perform our analysis, textual data needs to be transformed into numeric form. We call this process Feature Engineering. In this approach, numeric features are extracted or engineered from textual data. There are many Feature Engineering Techniques in existence, but for this problem we will use the Bag-of-Words model to engineer features.

We have already seen the process of cleaning data using feature extraction function from Scikit-Learn, and we will use these function again, which is repeative, but the best way to learn something is to do it more than once. In particular, we will use `TfidfTransformer()` and `CountVectorizer()` with the complaint data and the training and test sets we have aleady defined.

4.7.3 Vectorizing the Complaints

We have alerady have define `TfidfVectorizer()` to convert the collection of complaints (text) to a matrix of token counts, for which we can use in a Multinomial Naïve Bayes model as a classifier. The vectorizer we defined will transform the words into numbers representing the words. The output, text_bow_train and text_bow_text, will be used in the next step.

```
tfidf= TfidfVectorizer(sublinear_tf=True, min_df = 5, norm='l2',
encoding='latin-1', ngram_range=(1, 3), stop_words='english')
features = tfidf.fit_transform(X_train)
labels = y_train

vectorizer = CountVectorizer()
X_train = CountVectorizer().fit(X_train)
X_test = CountVectorizer().fit(X_test)

bow_transformer=vectorizer.fit(X_train)
text_bow_train=bow_transformer.transform(X_train) #TRAINING DATA
text_bow_test=bow_transformer.transform(X_test) #TEST DATA
```

4.8 BOW with a Multinomial Bayes Classifier

Finally, we define the `MultinomialNB()` to perform a fit of `X_train` and `y_train`, which we use as a classifier model. For simplicity, we will continue to use `CountVectorizer()`, but we'll name the output suing "bow" in the name. For example, the vectorizer with become `bow_transformer`, and the X_train data will become `text_bow_train`. Otherwise, the procedure is the same, except that we add so called **confusion matrices** to further visualize and analyze performance.

```
# instantiating the model with Multinomial Naive Bayes.
mnb_clf = MultinomialNB()
# training the model...
mnb_clf = mnb_clf.fit(text_bow_train, y_train)

print(mnb_clf)
MultinomialNB(alpha=1.0, class_prior=None, fit_prior=True)
```

4.8.1 Classifier Training Score

We now compute the `score()` for the trained classifier model. Score returns the mean accuracy on the given training or test data and labels. Yet, this is just a sanity check and not the score we want. We get the score of interest when we use the classifier to predict the predict group complaints using the two test sets, `X_test` and `y_test`.

```
mnb_clf.score(text_bow_train, y_train)
```

```
0.8205210096558185
```

4.8.2 Classifier Prediction

Now that we have the the fitted model, `mnb_clf`, we test its predictions, and measure its accuracy and performance. Notice that we only need to call the `mnb_clf` from the pipeline both to fit and predict. Then, we check the accuracy of our predictions using the the score module as we did above.

```
# Predict Outcomes using the MultinomialNB() model
y_pred = mnb_clf.predict(text_bow_test)
array(['credit_reporting', 'credit_card', 'credit_reporting','cr
edit_card', 'bank_service', 'debt_collection'], dtype='<U16')
```

4.8.3 Classifier Testing Score

Now, we score the model predictions for accuracy.

```
mnb_clf.score(text_bow_test, y_test)
```

```
0.8103683831383361
```

4.8.4 Confusion Matrices

Here we use what is called a "confusion matrix." A confusion matrix is a table that is often used to describe the performance of a classification model (or "classifier") on a set of test data for which the true values are known. It allows the visualization of the performance of an algorithm, which is the LOG in this case. The confusion matrix itself is relatively simple to understand, but the related terminology can be confusing.

You may have heard these terms before, perhaps in a probability and statistics course.

- True Positive: You predicted positive and it's true, i.e., you predicted that a complaint corresponds to a loan and it actually is.
- True Negative: You predicted negative and it's true, i.e., you predicted that a complaint does not corresponds to a loan and it actually is not.
- False Positive: (Type 1 Error) You predicted positive and it's false, i.e. you predicted that a complaint corresponds to a loan and it actually is not.
- False Negative: (Type 2 Error) You predicted negative and it's false. i.e. you predicted that a complaint does not correspond to a loan and it does.

Remember, we describe predicted values as Positive and Negative and actual values as True and False. The output below shows the confusion matrix, which is then plotted as a heatmap in Figure 4-11.

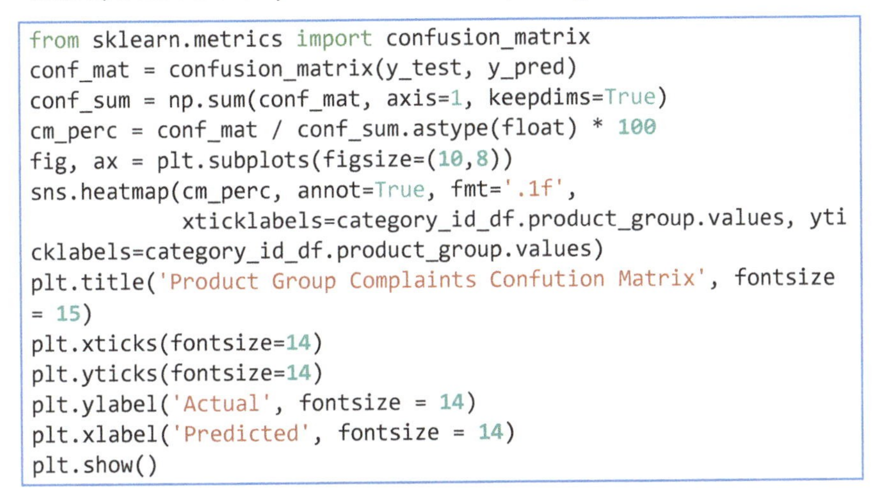

```
from sklearn.metrics import confusion_matrix
conf_mat = confusion_matrix(y_test, y_pred)
conf_sum = np.sum(conf_mat, axis=1, keepdims=True)
cm_perc = conf_mat / conf_sum.astype(float) * 100
fig, ax = plt.subplots(figsize=(10,8))
sns.heatmap(cm_perc, annot=True, fmt='.1f',
            xticklabels=category_id_df.product_group.values, yti
cklabels=category_id_df.product_group.values)
plt.title('Product Group Complaints Confution Matrix', fontsize
= 15)
plt.xticks(fontsize=14)
plt.yticks(fontsize=14)
plt.ylabel('Actual', fontsize = 14)
plt.xlabel('Predicted', fontsize = 14)
plt.show()
```

Figure 4-11. *Confusion matrix represented as a heatmap plot.*

4.8.5 Normalized Confusion Matrix

In Figure 4-12, we use a heatmap to plot a normalized confusion matrix for the bank product complaints.

```
from sklearn.metrics import accuracy_score, f1_score, precision_
score, recall_score, classification_report, confusion_matrix

result = mnb_clf.score(text_bow_test, y_test)
y_pred = mnb_clf.predict(text_bow_test)

cm_analysis(y_test, y_pred, mnb_clf.classes_, ymap=None, figsize
=(10,10))
```

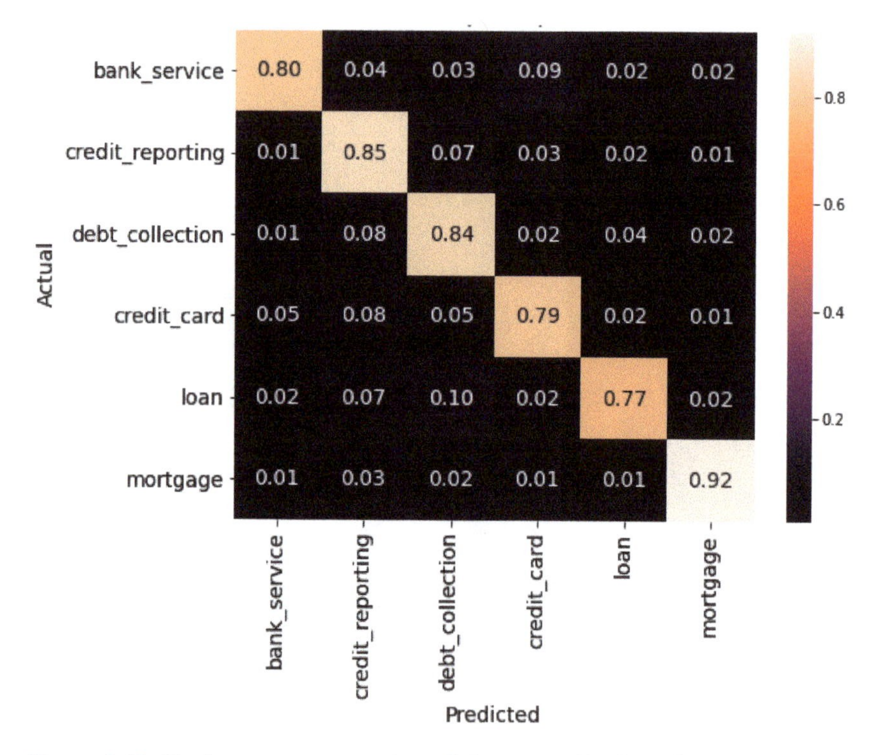

Figure 4-12. Heatmap representstion of the normalized confusion matrix for the bank product complaints.

4.8.6 Performance Metrics

4.8.7 Performance

Next, we conduct a more comprehensive performance test comprised of precision, recall, support, and f1-score metrics.

```
from sklearn import metrics
vectorizer = CountVectorizer()
y_test_counts = vectorizer.fit_transform(y_test)
features = vectorizer.get_feature_names()
print(metrics.classification_report(y_test, y_pred, target_names
=features))
```

```
                  precision    recall  f1-score   support

     bank_service       0.76      0.81      0.79      5999
      credit_card       0.72      0.76      0.74      8847
 credit_reporting       0.82      0.83      0.83     24445
  debt_collection       0.83      0.76      0.80     18481
             loan       0.74      0.78      0.76      9314
  money_transfers       0.86      0.63      0.73      1422
         mortgage       0.90      0.91      0.91     12006

         accuracy                           0.81     15819
        macro avg       0.81      0.81      0.81     15819
     weighted avg       0.82      0.81      0.81     15819
```

We combine this with the information provided by the confusion matrix, and conclude that the multinomial logistic classifier can be useful as a data science solution for accurate bank product group complaint classification. However, this was just an exploration to determine if further investigation is warranted. So, we now explore additional alternatives.

4.9 Classifier Comparison

In this section we will compare several models using accuracy as the metric for comparison. Then, we will take the best performing model and examine it in more detail as we did with the Naïve Bayes classifier. Th model we will run include:

- Multinomial Naïve Bayes Classifier

- Random Forest Calssifier
- Linear Support Vector Machine Classifier
- Ridge Regression Classifier
- Perceptron Classifier
- SGD Classifier
- Logistic Regression Classifier

```
from sklearn.feature_extraction.text import TfidfVectorizer
from sklearn.feature_extraction.text import HashingVectorizer
from sklearn.feature_selection import SelectFromModel
from sklearn.feature_selection import SelectKBest, chi2
from sklearn.linear_model import RidgeClassifier
from sklearn.svm import LinearSVC
from sklearn.linear_model import SGDClassifier
from sklearn.linear_model import Perceptron
from sklearn.linear_model import PassiveAggressiveClassifier
from sklearn.naive_bayes import BernoulliNB
from sklearn.naive_bayes import ComplementNB
from sklearn.naive_bayes import MultinomialNB
from sklearn.neighbors import KNeighborsClassifier
from sklearn.neighbors import NearestCentroid
from sklearn.ensemble import RandomForestClassifier
from sklearn.utils.extmath import density
from sklearn import metrics
```

4.9.1 Refresh Train and Test BOWs

An alternitve vectorizer, combining `CountVectorizer` and `TfidfTransformer`, is `TfidfVectorizer` (we used it for n-gram extraction).

```
tfidf= TfidfVectorizer(sublinear_tf=True, min_df=5, norm='l2', e
ncoding='latin-1', ngram_range=(1, 3), stop_words='english')
features = tfidf.fit_transform(X_train)
labels = y_train
```

```
tfidf= TfidfVectorizer(sublinear_tf=True, min_df = 5, norm='l2',
encoding='latin-1', ngram_range=(1, 3), stop_words='english')
features = tfidf.fit_transform(X_train)
labels = y_train

vectorizer = TfidfVectorizer()
X_train = TfidfVectorizer().fit(X_train)
X_test = TfidfVectorizer().fit(X_test)
```

```
bow_transformer=vectorizer.fit(X_train)
text_bow_train=bow_transformer.transform(X_train) #TRAINING DATA
text_bow_test=bow_transformer.transform(X_test) #TEST DATA
```

We want to take a seemingly redundant step and fresh the test and train sets that we transform into a bag-of-words, before we begin comparing classifiers.

```
# transforming into Bag-of-Words, hence textual data to numeric
text_bow_train=bow_transformer.transform(X_train) # TRAIN DATA
# transforming into Bag-of-Words, hence textual data to numeric
text_bow_test=bow_transformer.transform(X_test) # TEST DATA
```

4.9.2 Define Classifiers

First, we need to import the classifiers from Scikit-Learn. Next, we define models as being comprised of a the Linear Support Vector Machine Classifier (LinearSVC), the Stochastic Gradient Descent Classifier (SDGClassifier), the Multinomial Naïve Bayes (MultinomialNB) Classifier, the Ridge Classifier(RidgeClassifier), Perceptron Classifier, Multinomial Logistic Regression (LogisticRegression) Classifier.

```
from sklearn.linear_model import LogisticRegression
from sklearn.ensemble import RandomForestClassifier
from sklearn.svm import LinearSVC
from sklearn.model_selection import cross_val_score
features=text_bow_train
labels=y_train
models = [
          RandomForestClassifier(n_estimators=200, max_depth=3,
random_state=0),
          LinearSVC(),
          MultinomialNB(),
          SGDClassifier(),
          RidgeClassifier(tol=1e-2, solver="sag"),
          Perceptron(max_iter=25),
          LogisticRegression(random_state=0),
]
```

4.9.3 Execute the Classifiers

The code below will iterate (the "for" loop) through each classifier we defined in "models" above, fitting and scoring them for accuracy, and then plot them all on one accuracy plot for comparison. The plot is a combination of a boxplot and a stripplot. A strip plot is a scatterplot for categorical variables. It can be drawn on its own, but it a good complement to a box in cases where we want to show all observations along with some representation of the underlying distribution. Both plots are taken from the Seaborn module, and we abbreviate it as sns. From the plot, not that the Linear SVC, Multinomial Bayes, Ridge Regression and SGD Classifier, Perceptron, and Logistic Regression are in the range of good to strong classifiers in terms of accuracy. A Random Forest Classifier performs poorly with this data. Figure 4-13 shows the performance of the six slected classifiers, using a combination scatterplot and box plots.

```
import seaborn as sns
import warnings
warnings.filterwarnings("ignore",category=DeprecationWarning)

CV = 5
cv_df = pd.DataFrame(index=range(CV * len(models)))
entries = []
for model in models:
  model_name = model.__class__.__name__
  accuracies = cross_val_score(
model, features, labels, scoring='accuracy', cv=CV)
for fold_idx, accuracy in enumerate(accuracies):
entries.append((model_name, fold_idx, accuracy))
cv_df = pd.DataFrame(entries,
columns=['model_name', 'fold_idx', 'accuracy'])
import seaborn as sns
plt.figure(figsize=[12,6])
sns.boxplot(x='model_name', y='accuracy', data=cv_df)
sns.stripplot(x='model_name', y='accuracy', data=cv_df,
hue='model_name', size=12, jitter=True,
edgecolor="gray", linewidth=2)
plt.title('Classifier Accuracy Comparison', fontsize=14)
plt.ylabel('Accuracy', fontsize=14)
plt.xlabel('Classifier Name', fontsize=14)
plt.xticks(fontsize=12)
plt.yticks(fontsize=14)
plt.show()
```

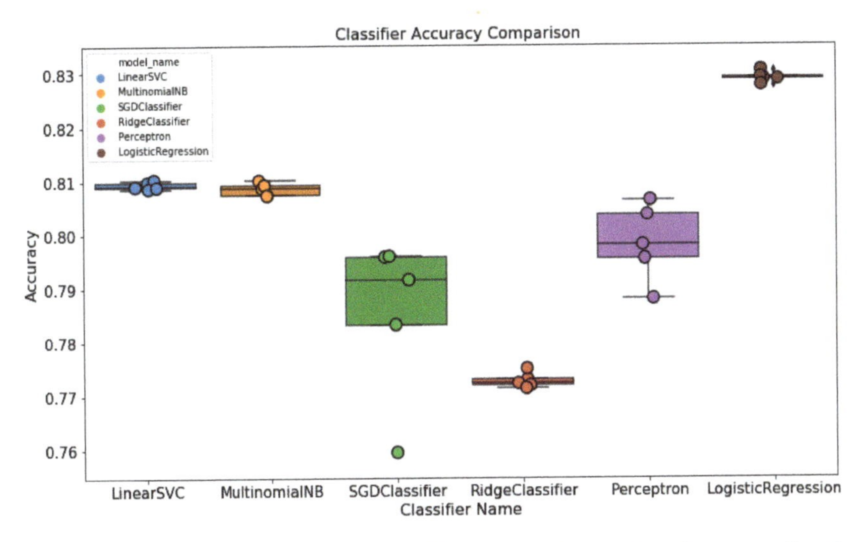

Figure 4-13. Combined scatterplot and box plots of model performance for six classifiers (from left to right: linear SVC, multinomial naïve Bayes, SGD classifier, ridge classsifier, perceptron, and logistic regression).

4.10 Model Analysis

Now we turn to our analysis of the models output and performance.

4.10.1 Anlysis of Model Performance

To start with, we can see on the graph above our models from left to right: the LinearSVC, the MultinomialNB classifier, the SDGClassifier, the RidgeClassifier, Perceptron classifier, LogisticRegression classifier. From the graph it appears that the multinomial logistic regression clasifier provides the best performance. This is supported by the model accuracy metric shown below, with about 83%.

```
cv_df.groupby('model_name').accuracy.mean()
```

```
model_name
LinearSVC             0.809242
LogisticRegression    0.833755
MultinomialNB         0.808465
Perceptron            0.798186
RidgeClassifier       0.772593
SGDClassifier         0.785113
Name: accuracy, dtype: float64
```

4.10.2 BOW Analysis with a Multinomial Logistic Regression

Having concluded that the Multinomial Logistic Regression is the best classifier, we now implement it shoing the full details and analysis, in the manner we first did with the Multinomial Naïve Bayes Classifier. So, we first fit the data with the classifier as we dd before, we use a pipeline to prepare the data and to define the model

```
from sklearn.linear_model import LogisticRegression

log_clf = Pipeline([
('vect', vectorizer),
('tfidf', TfidfTransformer()),
('svm_clf', LogisticRegression ()),
])
log_clf = LogisticRegression ()
log_clf.fit(text_bow_train, y_train)
```

```
LogisticRegression (C=1.0, class_weight=None, dual=True,
   fit_intercept=True,intercept_scaling=1, loss='squared_hinge',
   max_iter=1000,multi_class='ovr', penalty='l2',
   random_state=None, tol=0.0001, verbose=0)
```

Here, we get an intial glimps at the classifier's performance score.

```
log_score = log_clf.score(text_bow_train, y_train)
log_score
```

```
0.8783547847934166
```

Next, we use the classier to run the test data and look at the intial performace information, which tells us that the model performes well, even though the score differ by about 0.04.

```
y_pred = log_clf.predict(text_bow_test)
y_pred_score = log_clf.score(text_bow_test,y_test)
y_pred_score
```

```
0.8337556201405967
```

4.10.3 Confusion Matrix

In Section 4.8.6, we discussed how to create and interpret a confusion matrix for a multinomial Bayes classifier model. Here we generate the confusion matrix for the linear support vector machine model. The mean value for the *logistic regression* is better than that of the NBM, so we go an extra step in measuring its performance. Here we use what is called a "confusion matrix." A confusion matrix is a table that is often used to describe the performance of a classification model (or "classifier") on a set of test data for which the true values are known. It allows the visualization of the performance of an algorithm, which is the *logistic regression* in this case. The confusion matrix itself is relatively simple to understand, but the related terminology can be confusing.

We defined the terminaolgy in Section 4.8.5 but repeat it here for convenience.

1. True Positive: You predicted positive and it's true, i.e., you predicted that a complaint corresponds to a loan and it actually is.
2. True Negative: You predicted negative and it's true, i.e., you predicted that a complaint does not corresponds to a loan and it actually is not.
3. False Positive: (Type 1 Error) You predicted positive and it's false, i.e. you predicted that a complaint corresponds to a loan and it actually is not.
4. False Negative: (Type 2 Error) You predicted negative and it's false. i.e. you predicted that a complaint does not correspond to a loan and it does.

Remember, we describe predicted values as Positive and Negative and actual values as True and False. The results are shown in .

```
from sklearn.metrics import confusion_matrix
conf_mat = confusion_matrix(y_test, y_pred)
fig, ax = plt.subplots(figsize=(10,10))
plt.xticks(fontsize=12)
plt.yticks(fontsize=12)
sns.heatmap(conf_mat, annot=True, fmt='d',
            xticklabels=category_id_df.product_group.values, yti
cklabels=category_id_df.product_group.values)
plt.title('Product Group Complaints Confution Matrix', fontsize
= 15)
plt.ylabel('Actual', fontsize = 12)
plt.xlabel('Predicted', fontsize = 12)
plt.show()
```

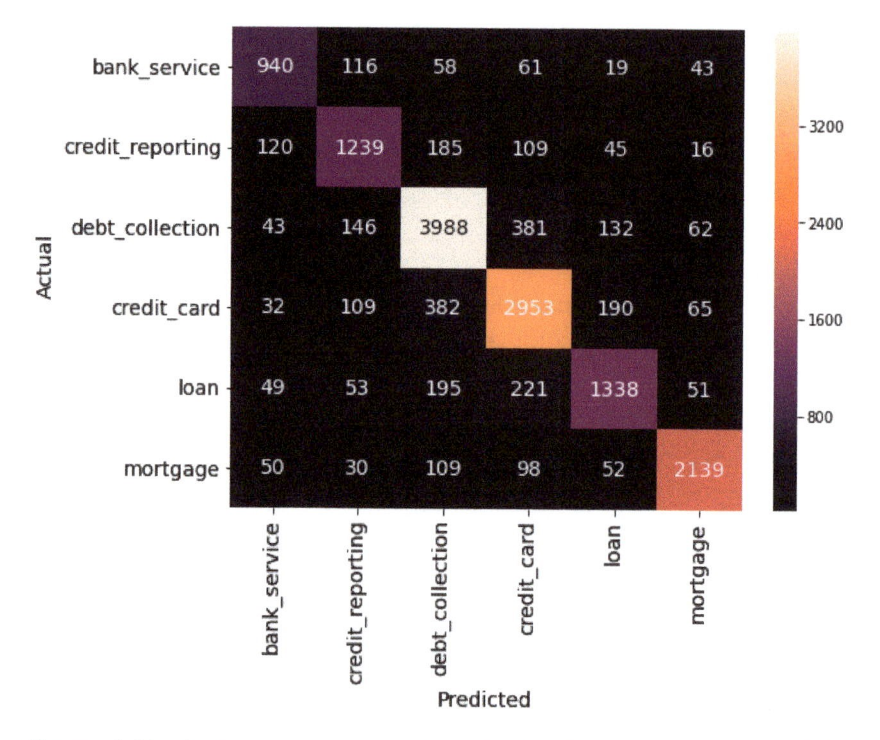

Figure 4-14. Heatmap representation of the bank products complaint confusion matrix.

4.10.4 Normalized Confusion Matrix

Given the matrix below and 96.653% accuracy:

- 'Bank Service' is accurately 73.2% of the time, and incorrectly classifies as a 'credit card' complaint 9.2% of the time, as a 'credit reporting' complaint 4.7% of the time, and so on, as we move left to right on row one.
- 'Credit Card' is accurately 71.1% of the time, and is incorrectly classified as a 'credit reporting' complaint 11.2% of the time, as a 'debt collection' complaint 6.97 of the time, so on.
- 'Credit Reporting' is accurately 87.3% of the time, and incorrectly classified as a 'debt collection, complaint 6.2% of the time, as a 'loan' complaint 2.3% of the time, and so on.
- 'Debt Collection' is accurately 81.5% of the time, and incorrectly classied as a 'loan' 3.8% of the time, as a 'money transfer' complaint 0.1% of the time, and so on.

- 'Loan' is accurately 73.8% of the time, and incorrectly classied as a 'money transfer' 0.3% of the time, as a 'mortgage' complaint 2.4% of the time, and so on.
- 'Money Transfers' is accurately 68.6% of the time, and incorrectly classified as a 'mortgage' complaint 1.7% of the time, as a 'bank service' complaint 14.9% of the time and so on.
- 'Mortgages' is accurately 88.9% of the time, and incorrectly classified as a 'bank service' complaint 1.4% of the time, as a 'credit card' pmplaint 0.7%" of the time, and so on..

Accuracy - 81.5%, Recall = 77.7%, Precision is 79.6%, and the F1 Score is 78.6%

```
from sklearn.metrics import confusion_matrix
conf_mat = confusion_matrix(y_test, y_pred)
conf_sum = np.sum(conf_mat, axis=1, keepdims=True)
cm_perc = conf_mat / conf_sum.astype(float) * 100
fig, ax = plt.subplots(figsize=(10,10))
plt.xticks(fontsize=12)
plt.yticks(fontsize=12)
sns.heatmap(cm_perc, annot=True, fmt='.1f',
            xticklabels=category_id_df.product_group.values, yti
cklabels=category_id_df.product_group.values)

plt.title('Normalized Product Group Complaints Confution Matrix'
, fontsize = 15)
plt.ylabel('Actual', fontsize = 12)
plt.xlabel('Predicted', fontsize = 12)
plt.show()
```

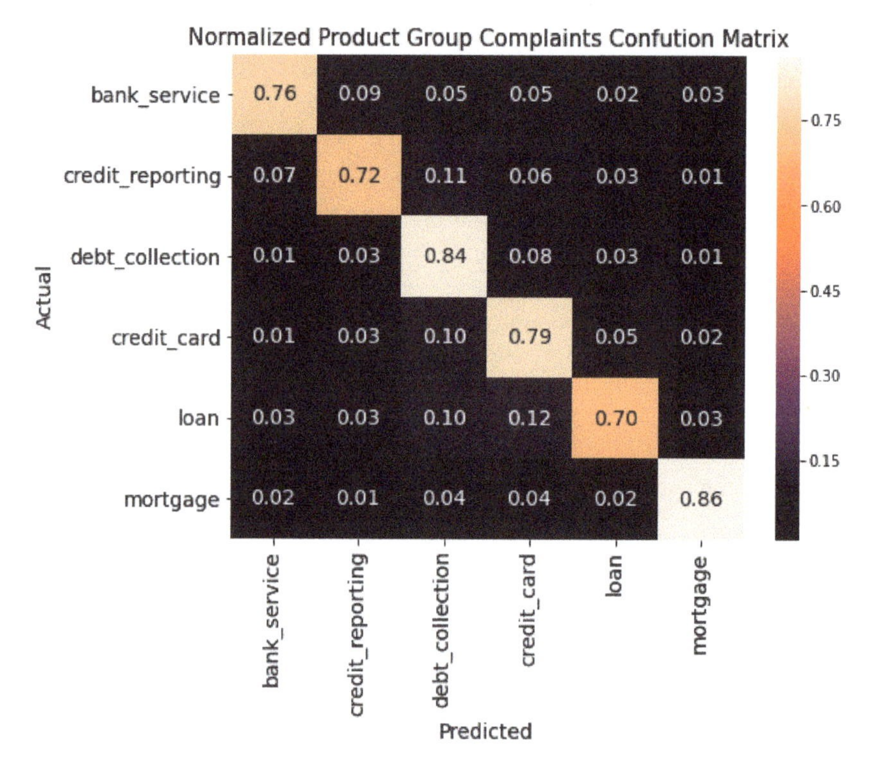

Figure 4-15. *Heatmap representation of the bank products complaint normalized confusion matrix.*

Finally, we run a complete evaluation of the classifiers performance by examining its accuracy, F1 Score, precision score, and recall score. We explain these in subsequent sections.

```
result = log_clf.score(text_bow_test, y_test)
print("Accuracy: %.3f%%" % (result*100.0))
y_pred = log_clf.predict(text_bow_test)
print("F1 Score: ", metrics.f1_score(y_test, y_pred,
        average="macro"))
print("Precision Score: ", metrics.precision_score(y_test,
        y_pred, average="macro"))
print("Recall Score: ", metrics.recall_score(y_test, y_pred,
        average="macro"))
```

```
Accuracy: 83.376%
F1 Score:  0.8113303656236959
Precision Score:  0.8302707674566788
Recall Score:  0.7956411102363623
```

4.10.5 Classification Report (Precision, Recall and F1–Score)

We just looked at the overall performance records for the Logistic Regression Classifier. Now we turn to the performance metrics as they pertain to each product group.

```
print(metrics.classification_report(y_test,y_pred))
```

```
                   precision    recall  f1-score   support

     bank_service       0.76      0.76      0.76      1237
      credit_card       0.73      0.72      0.73      1714
 credit_reporting       0.81      0.84      0.82      4752
  debt_collection       0.77      0.79      0.78      3731
             loan       0.75      0.70      0.73      1907
         mortgage       0.90      0.86      0.88      2478

         accuracy                           0.80     15819
        macro avg       0.79      0.78      0.78     15819
     weighted avg       0.80      0.80      0.80     15819
```

4.10.6 Conclusion

The Logistic Regression classifier exhibits the most accuracy of all the fitted models. However, the other models we fitted are also accurate within five percentage points, the least accurate being the Ridge Classifier.

4.11 Metrics Defined

We briefly touched metrics in Section 4.8.4 (Performace). Now, lets peel back the onion and look a little deeper.

4.11.1 Accuracy

Accuracy, as we saw earlier, is a metriic often used to evaluate the performance of a classifer. For example, we saw that the accuracy of the logistic regression model is 0. 828809. However, it may not be the right measure at times, especially if our Target class is is skewed (left or right). Then we may want to consider other metrics like Precision, Recall, F-score (combined metric), which we have yet to define (Rite, 2018).

4.11.2 Precision

Precision is the “exactness” or the ability of our clasifier to return only relevant instances. If our problem statement or use case involves minimizing the False Positives, i.e. in current scenario if you don’t want the *loan compalints* to be labelled as *credit card complaints* by the model then Precision is something we need.

Definition 4.4. *Presicion is defined as the number of true positives (TP) divided by the sum of the true positive and false positives (FP), or*

$$\frac{TP}{TP + FP}$$

For our situation, we have 6,928 TP for loan complaints and 1368 that are misclassified as other product complaints.

```
# Precision
Precision = tp/(tp+fp)
print("Precision {:0.2f}".format(Precision))
```

```
Precision 0.8351012536
```

So, in terms of precision, the logistic regression classifier is on par with the accuracy of the classifier, 0.83376.

4.11.3 Recall

Recall is the “completeness”, or the ability of the model to identify all relevant instances, True Positive Rate, or *sensitivity*. In the current scenario if our focus is to have the least False Negatives i.e., we don’t laon complaints to be **wrongly** classified as credit complaints then Recall is what we may need.

Definition 4.5. *Recall is defined as the number of true positives (TP) divided by the sum of the true positive and false negatives (FN), or*

$$\frac{TP}{TP + FN}$$

For our situation, we have 6,928 TP for loan complaints and 1431 that are wrongly classified as loan complaints.

```
# Recall
Recall = tp/(tp+fn)
print("Recall {:0.2f}".format(Recall))
```

```
Recall: 0.8288072736
```

So, in terms of recall, the logistic regression classifier is on par with the accuracy of the classifier, 0.83376.

4.11.4 F1 Measure

The F1 measure is the harmonic mean of *precision* and *recall*, used to indicate a balance between precision and recall providing each is equally weighted, and ranges from 0 to 1. F1 measure reaches its best value at 1 (perfect precision and recall) and worst at 0.

Definition 4.6. *The harmonic mean of precision and recall is:*

$$\frac{2 \times \boldsymbol{precision} \times \boldsymbol{recall}}{\boldsymbol{precsion} + \boldsymbol{recal}}$$

In our situation we have:

$$\frac{2 \times 0.8351012536 \times 0.8288072736}{0.8351012536 + 0.8288072736} = 0.83194236$$

```
# F1 Score
f1 = (2*Precision*Recall)/(Precision + Recall)
print("F1 Score {:0.2f}".format(f1))
```

```
f1: 83194236
```

These calculations are consistent with our previous findings.

4.12 Ridge Regression Classifier

Now, let's look at these measures for the Ridge Classifier. Recall the accuracy for this model was lower than all the others.

```
rg_clf = RidgeClassifier(tol=1e-2, solver="sag")
rg_clf.fit(text_bow_train, y_train)
rg_score = rg_clf.score(text_bow_train, y_train)
rg_score
```

```
0.7787731681091842
```

```
y_pred = rg_clf.predict(text_bow_test)
y_pred_score = rg_clf.score(text_bow_test, y_test)
y_pred_score
```

```
0.7732071440991629
```

We show the confussion matric for the ridge classifiers in **Figure 4-16**. Notice that even though the classifer shows 77% accuracy, several of the bank product complaints result in accurate predictions lower than 70%.

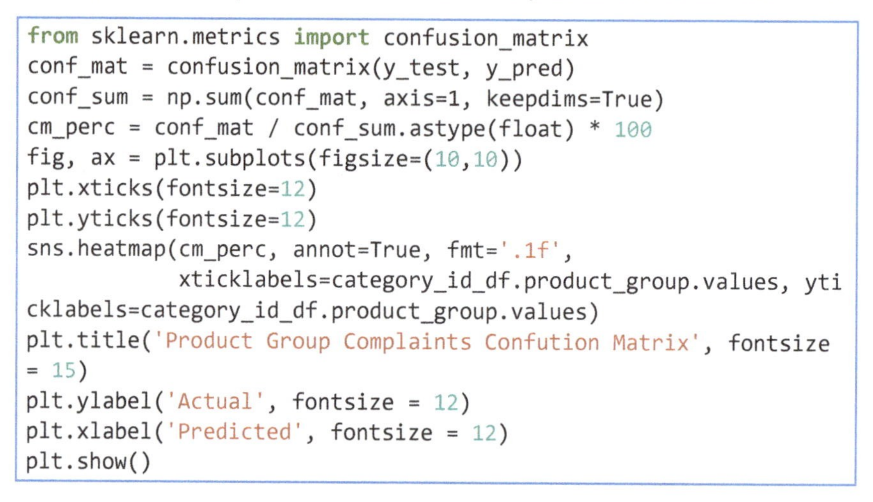

```
from sklearn.metrics import confusion_matrix
conf_mat = confusion_matrix(y_test, y_pred)
conf_sum = np.sum(conf_mat, axis=1, keepdims=True)
cm_perc = conf_mat / conf_sum.astype(float) * 100
fig, ax = plt.subplots(figsize=(10,10))
plt.xticks(fontsize=12)
plt.yticks(fontsize=12)
sns.heatmap(cm_perc, annot=True, fmt='.1f',
            xticklabels=category_id_df.product_group.values, yti
cklabels=category_id_df.product_group.values)
plt.title('Product Group Complaints Confution Matrix', fontsize
= 15)
plt.ylabel('Actual', fontsize = 12)
plt.xlabel('Predicted', fontsize = 12)
plt.show()
```

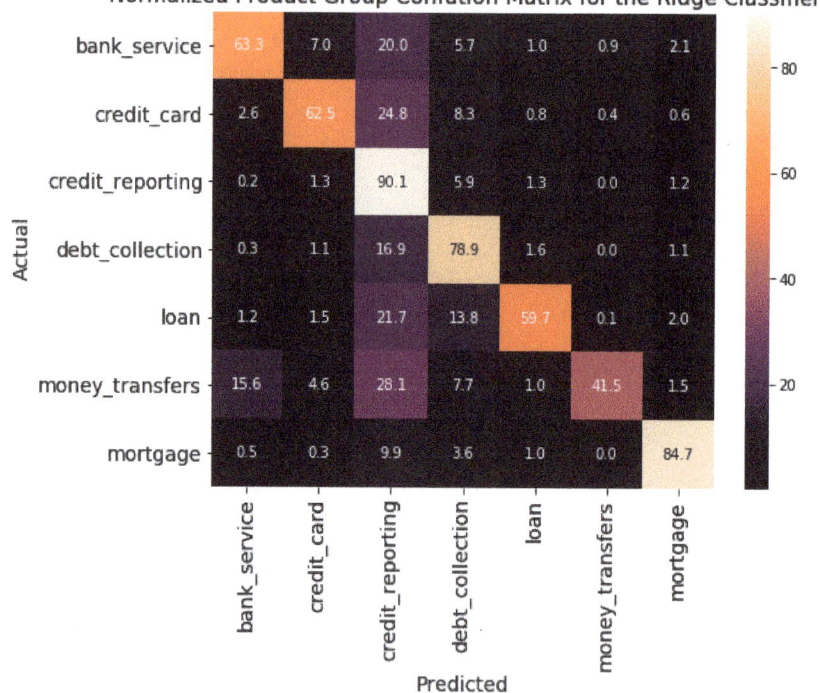

__Figure 4-16.__ Heatmap representation of the bank products complaint confusion matrix for the ridge classifier.

```
result = rg_clf.score(text_bow_test, y_test)
print("Accuracy: %.3f%%" % (result*100.0))
y_pred = rg_clf.predict(text_bow_test)
print("F1 Score: ", f1_score(y_test, y_pred, average="macro"))
print("Precision Score: ", precision_score(y_test, y_pred, avera
ge="macro"))
print("Recall Score: ", recall_score(y_test, y_pred, average="ma
cro"))
```

```
Accuracy: 77.329%
F1 Score:  0.733543853625606
Precision Score:  0.819857261962727
Recall Score:  0.686976921614076
```

```
print(classification_report(y_test,y_pred))
```

```
                    precision    recall   f1-score    support
bank_service             0.84      0.63      0.72       5999
credit_card              0.83      0.63      0.71       8847
credit_reporting         0.68      0.90      0.78      24445
debt_collection          0.77      0.79      0.78      18481
loan                     0.86      0.60      0.71       9314
money_transfers          0.84      0.41      0.55       1422
mortgage                 0.92      0.85      0.88      12006
avg / total              0.82      0.77      0.77      80514
```

So, in terms of precision, for example, the ridge regression classifier is as good as the logistic regression classifier. We will leave it to an exercise to compare the precsions of the remaining classifiers:

- LinearSVC
- MultinomialNB
- Perceptron
- SGDClassifier

4.13 Exercises

1. For bank product complaint classifiers, compare these four classifers (LinearSVC, MultinomialNB, Perceptron, and SGDC Classifier) in terms of *precision* with the logistic regression classifier. State your conclusions.
2. Repeat Excerise 1 for *Recall*.
3. Repeat Excerise 1 for *F1*.
4. The following confusion matrix represents violations of thr Servicemembers Civil Relief Act (SCRA) in banking, divided into five categories or features: 'Checking or savings account','Consumer Loan','Credit card','Debt collection','Mortgage'. What can be done to make more precise classifier?

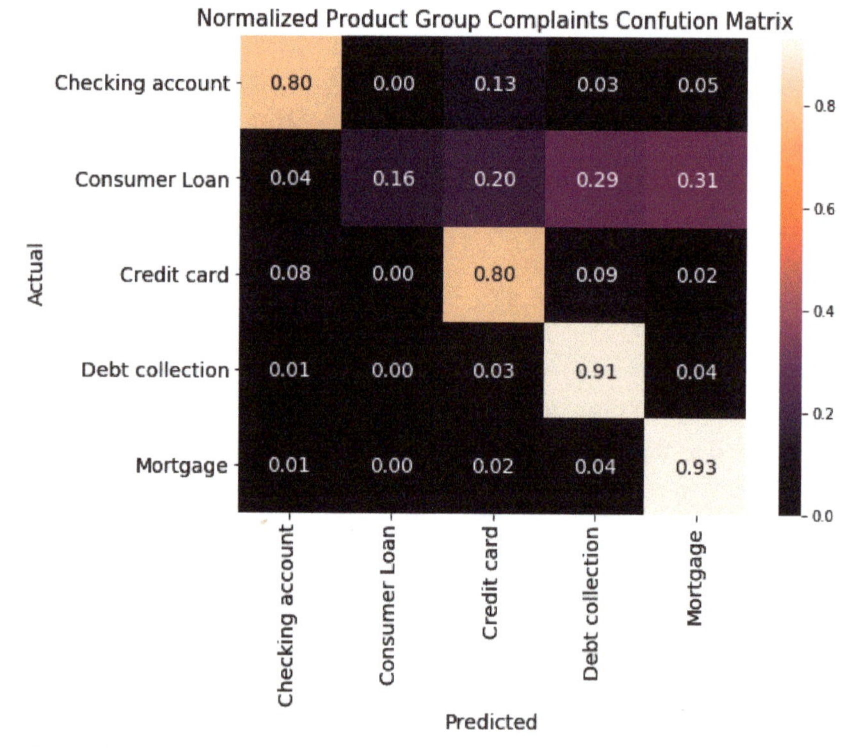

5. Mimic the example in **Section 4.12** to classify a bank's violation of the Servicemembers Civil Relief Act (SCRA) using the data at: *https://github.com/stricje1/Data/blob/master/scra_complaint_data.csv*. [Hint: the heatmap in Excersize 4 represents a solution.]

6. Given the SCRA data from Exercise 5, use Multi-Layer Perceptron (MLP) classifier and a bag-of-words technique to classify the SCRA violations (by product). Does the algorithm converge? If it converges, at what step does it converge and what is the score at convergence?
7. Using the SCRA data from Exercise 5, find the bigrams (see Section 4.6.5.

Chapter 5 – Tweet Sentiment Analysis

5.1 Introduction

Twitter is now a mainstream social media network. The streams of conversation include politics, like President Trup and ISIS Fanbos, business, stars, celebrities, and so on. Suppose we could make sense of all the Tweets from a sigle author or group. We can do this through sentiment analysis.

Definition 5.1. *A sentiment analysis system for text analysis combines NLP and machine learning techniques to assign weighted sentiment scores to the entities, topics, themes and categories within a sentence or phrase.*

5.2 Sentiment Lexicons

One way to analyze the sentiment of a text is to consider the text as a combination of its individual words and the sentiment content of the whole text as the sum of the sentiment content of the individual words. This is often performed using lexicons. Linguistic theories generally regard human languages as consisting of two parts: a lexicon, essentially a catalogue of a language's words (its wordstock); and a grammar, a system of rules which allow for the combination of those words into meaningful sentences.

Definition 5.2. *A lexicon is a dictionary that includes or focuses on lexemes, a unit of lexical meaning, roughly corresponding to the set of inflected forms taken by a single word. For example, the lexeme run includes as members "run" (lemma), "running" (inflected form), and "ran", but excludes "runner" (a derived term).*

Lexicons used of sentiment analysis contain sentiment words, like "trust" and "disgust," and all the words associated with a particular sentiment. These lexicons are "lined" togther with the text under analysis by performing an inner-join with the text, the details of which we will discuss later.

Sentiment lexicons constructed via either crowdsourcing (using, for example, Amazon Mechanical Turk) or by the labor of one of the authors, and are validated using some combination of crowdsourcing again, restaurant or movie reviews, or Twitter data. Given this information, we may hesitate to apply these sentiment lexicons to styles of text dramatically different from what they were validated on, such as narrative fiction from 200 years ago. While it is true that using these sentiment lexicons with "Indian Philosophy," for example, may give us less accurate results than with tweets sent by a contemporary writer, we still can measure the sentiment content for words that are shared across the lexicon and the text. The three general-purpose lexicons are - AFINN from Finn Årup Nielsen, - bing from Bing Liu and collaborators, and – Loughran from Tim Loughran and Bill McDonald.

All three of these lexicons are based on unigrams, i.e., single words. These lexicons contain many English words and the words are assigned scores for positive/negative sentiment, and also possibly emotions like joy, anger, sadness, and so forth. The loughran lexicon categorizes words in a binary fashion ("yes"/"no") into categories of positive, negative, anger, anticipation, disgust, fear, joy, sadness, surprise, and trust. The bing lexicon categorizes words in a binary fashion into positive and negative categories. The AFINN lexicon assigns words with a score that runs between -5 and 5, with negative scores indicating negative sentiment and positive scores indicating positive sentiment. All of this information is tabulated in the sentiments dataset, and tidytext provides a function `get_sentiments()` to get specific sentiment lexicons without the columns that are not used in that lexicon.

Not every English word is in the lexicons because many English words are pretty neutral. It is important to keep in mind that these methods do not consider qualifiers before a word, such as in "no good" or "not true"; a lexicon-based method like this is based on unigrams only.

5.3 Install Required Libraries

Most of the R packages we need will be loaded here (and installed is needed).

```
if(!require(tidytext)) install.packages("tidytext")
if(!require(tidyverse))install.packages("tidyverse")
if(!require(dplyr)) install.packages("dplyr")
if(!require(tidyr)) install.packages("tidyr")
if(!require(sentimentr))install.packages("sentiment")
if(!require(tm)) install.packages("tm")
if(!require(readr)) install.packages("readr")
```

```
if(!require(wordcloud))install.packages("wordcloud")
if(!require(lubridate))install.packages("lubridate")
if(!require(ggplot2)) install.packages("ggplot2")
if(!require(ggraph)) install.packages("ggraph")
if(!require(igraph)) install.packages("igraph")
if(!require(plotrix)) install.packages("plotrix")
```

5.4 Lexicon Example

The *tidytext* package contains several sentiment lexicons in the sentiments dataset.

```
library(tidytext)
sentiment
```

We can look at the way specific lexicons score sentiment. The Bing lexicon categorizes words in a binary fashion into positive and negative categories.

```
get_sentiments("bing")
```

```
# A tibble: 6,788 x 2
   word        sentiment
   <chr>       <chr>
 1 2-faced     negative
 2 2-faces     negative
 3 a+          positive
 4 abnormal    negative
 5 abolish     negative
 6 abominable  negative
 7 abominably  negative
 8 abominate   negative
 9 abomination negative
10 abort       negative
# ... with 6,778 more rows
```

The AFINN lexicon assigns words with a score that runs between -5 and 5, with negative scores indicating negative sentiment and positive scores indicating positive sentiment.

```
get_sentiments("afinn")
```

```
# A tibble: 2,476 x 2
   word       score
   <chr>      <int>
 1 abandon       -2
 2 abandoned     -2
 3 abandons      -2
 4 abducted      -2
```

```
  5 abduction     -2
  6 abductions    -2
  7 abhor         -3
  8 abhorred      -3
  9 abhorrent     -3
 10 abhors        -3
# ... with 2,466 more rows
```

The Loughran lexicon categorizes words in a binary fashion ("yes"/"no") into categories of positive, negative, anger, anticipation, disgust, fear, joy, sadness, surprise, and trust.

```
get_sentiments("loughran")
```

```
# A tibble: 13,901 x 2
   word        sentiment
   <chr>       <chr>
 1 abacus      trust
 2 abandon     fear
 3 abandon     negative
 4 abandon     sadness
 5 abandoned   anger
 6 abandoned   fear
 7 abandoned   negative
 8 abandoned   sadness
 9 abandonment anger
10 abandonment fear
# ... with 13,891 more rows
```

5.5 Example 1: The Inner Join

With data in a tidy format, sentiment analysis can be done as an inner-join. This is another of the great successes of viewing text mining as a tidy data analysis task; much as removing stop words is an anti-join operation, performing sentiment analysis is an inner-join operation. For this example, we will use a collection of tweets from President Donald Trump, as the lexicons are geared more toward the analysis of tweets than the analysis of ancient philosophies.

```
setwd("C:/Users/jeff/Documents/VIT_Course_Material/Data_Analytic
s_2018/data/")
trump_tweets<-read.csv("trump_tweets.csv",stringsAsFactor=FALSE)
str(trump_tweets)
```

```
'data.frame':    1050 obs. of  2 variables:
 $ time  : chr  "8/13/2018 8:57" "8/13/2018 9:21" "8/11/2018 1:
28" "8/8/2018 10:14" ...
 $ tweets: chr  " ....such wonderful and powerful things about
me - a true Champion of Civil Rights - until she got fired. Omar
```

```
o"| __truncated__ " While I know it's "not presidential" to take
on a lowlife like Omarosa, and while I would rather not be doing
"| __truncated__ " The big story that the Fake News Media refuse
s to report is lowlife Christopher Steele's many meetings with D
e"| __truncated__ " The Republicans have now won 8 out of 9 Hous
e Seats, yet if you listen to the Fake News Media you would thin
k "| __truncated__ ...
```

Observe that trump_tweets is alerady a data frame, and that there are 1050 tweets in the dataset, with line one as a header. We will look at the words with a "trust" score from the `loughran` lexicon. What are the most common "trust" words in Trump_Tweets? First, we need to take the text of the novels and convert the text to the tidy format using `unnest_tokens()`.

The next code chunk converts the time to date in the dataset, which will be ordered by date.

```
trump_tweets$date<-as.Date(trump_tweets$time, "%m/%d/%Y %H:%M")
head(trump_tweets)
```

```
              time
 1 8/13/2018 8:57
 2 8/13/2018 9:21
 3 8/11/2018 1:28
 4 8/8/2018 10:14
 5  8/5/2018 7:49

tweets
 1  ....such wonderful and powerful things about me - a true Cha
mpion of Civil Rights - until she got fired. Omarosa had Zero cr
edibility with the Media (they didn't want interviews) when she
worked in the White House. Now that she says bad about me, they
will talk to her. Fake News! [Twitter for iPhone] link
 2  While I know it's "not presidential" to take on a lowlife l
ike Omarosa, and while I would rather not be doing so, this is a
modern-day form of communication and I know the Fake News Media
will be working overtime to make even Wacky Omarosa look legitim
ate as possible. Sorry! [Twitter for iPhone] link
 3  The big story that the Fake News Media refuses to report is
lowlife Christopher Steele's many meetings with Deputy A.G. Bruc
e Ohr and his beautiful wife, Nelly. It was Fusion GPS that hire
d Steele to write the phony & discredited Dossier, paid for by C
rooked Hillary & the DNC.... [Twitter for iPhone] link
 4  The Republicans have now won 8 out of 9 House Seats, yet if
you listen to the Fake News Media you would think we are being c
lobbered. Why can't they play it straight, so unfair to the Repu
blican Party and in particular, your favorite President! [Twitte
r for iPhone] link
```

```
 5  Fake News reporting, a complete fabrication, that I am conce
rned about the meeting my wonderful son, Donald, had in Trump To
wer. This was a meeting to get information on an opponent, total
ly legal and done all the time in politics - and it went nowhere
. I did not know about it! [Twitter for iPhone] link
        date
 1 2018-08-13
 2 2018-08-13
 3 2018-08-11
 4 2018-08-08
 5 2018-08-05
```

Next, we will tokenize the text.

```
trump_tweets<-trump_tweets %>% group_by(date) %>%
mutate(ln=row_number())%>%
unnest_tokens(word,tweets) %>% ungroup()
head(trump_tweets,4)
```

```
 # A tibble: 5 x 4
   time              date          ln word
   <chr>             <date>     <int> <chr>
 1 8/13/2018 8:57 2018-08-13       1 such
 2 8/13/2018 8:57 2018-08-13       1 wonderful
 3 8/13/2018 8:57 2018-08-13       1 and
 4 8/13/2018 8:57 2018-08-13       1 powerful
```

Notice that we chose the name word for the output column from **unnest_tokens()**. This makes performing inner joins and anti-joins is thus easier because the sentiment lexicons and stop word datasets have columns named word.

5.5.1 Using the Inner_Join to Analyze Sentiment

Now that the text is in a tidy format with one word per row, we are ready to do the sentiment analysis. First, let's use the **loughran** lexicon and **filter()** for the "trust" words. Next, we will **filter()** the data frame with the text from the trump_tweets for the words and then use **inner_join()** to perform the sentiment analysis. What are the most common "trust" words in **trump_tweets**? We'll use **count()** from *dpLyr* get this answer.

```
loughran _trust <- get_sentiments("Loughran") %>%
  filter(sentiment == "trust")
trump_tweets %>%
  inner_join(loughran _trust) %>%
  count(word, sort = TRUE)
```

```
Joining, by = "word"
# A tibble: 189 x 2
   word            n
   <chr>       <int>
 1 president      63
 2 show           31
 3 enjoy          25
 4 good           20
 5 vote           20
 6 real           19
 7 credibility    18
 8 money          18
 9 trade          17
10 white          17
# ... with 179 more rows
```

We see mostly positive words associated with "trust". Now we'll look and see what "disgust" the President has.

```
loughran_disgust <- get_sentiments("loughran") %>%
  filter(sentiment == "disgust")
trump_tweets %>%
  inner_join(loughran_disgust) %>%
  count(word, sort = TRUE)
```

```
Joining, by = "word"
# A tibble: 126 x 2
   word          n
   <chr>     <int>
 1 bad          66
 2 dishonest    49
 3 phony        25
 4 witch        21
 5 dying        19
 6 collusion    18
 7 enemy        17
 8 john         16
 9 terrible     12
10 hate         11
# ... with 116 more rows
```

We can also examine how sentiment changes throughout the time period (10/19/2011 to 8/13/2018). We can do this with just a handful of lines that are mostly *dplyr* functions. First, we find a sentiment score for each word using the Bing lexicon and `inner_join()`. Next, we count up how many positive and negative words there are in defined in each Tweet. We then use `spread()` so that we have negative and positive sentiment in separate columns, and lastly calculate a net sentiment (positive - negative).

```
library(tidyr)
trump_sentiment <- trump_tweets %>% inner_join(get_sentiments("b
ing"))
```

```
Joining, by = "word"
```

Notice that we are plotting against the index on the x-axis that keeps track of narrative time in sections of text. Using ggplot2, the plot is displayed in Figure 5-1, using the code below.

```
library(ggplot2)
trump_tweets%>%inner_join(get_sentiments("Loughran")) %>%count(w
ord,sentiment) %>% group_by(sentiment)%>%top_n(10)%>%
  ungroup() %>%mutate(word=reorder(word,n))%>%ggplot(aes(x=word,
y=n,fill=sentiment)) +
  geom_col(show.legend = FALSE) +
  facet_wrap(~ sentiment, scales = "free") +
  coord_flip()
```

5.5.2 Positive & Negative Words over time

We can also look at the sentiment over time, like a time series, and note the trends, sesonality, noise, and so on. Below is the code for sentiment by time, which is displayed in Figure 5-2. Notice the interpolated lines (dashed) have a negative slope, with negative sentiment decreasing at a fater rate.

```
sentiment_by_time <- trump_tweets %>%
  mutate(dt = floor_date(date, unit = "month")) %>%
  group_by(dt) %>%
  mutate(total_words = n()) %>%
  ungroup() %>%
  inner_join(get_sentiments("loughran"))
```

```
Joining, by = "word"
```

```
sentiment_by_time %>%
  filter(sentiment %in% c('positive','negative')) %>%
  count(dt,sentiment,total_words) %>%
  ungroup() %>%
  mutate(percent = n / total_words) %>%
  ggplot(aes(x=dt,y=percent,col=sentiment,group=sentiment)) +
  geom_line(size = 1.5) +
  geom_smooth(method = "lm", se = FALSE, lty = 2) +
  expand_limits(y = 0)
```

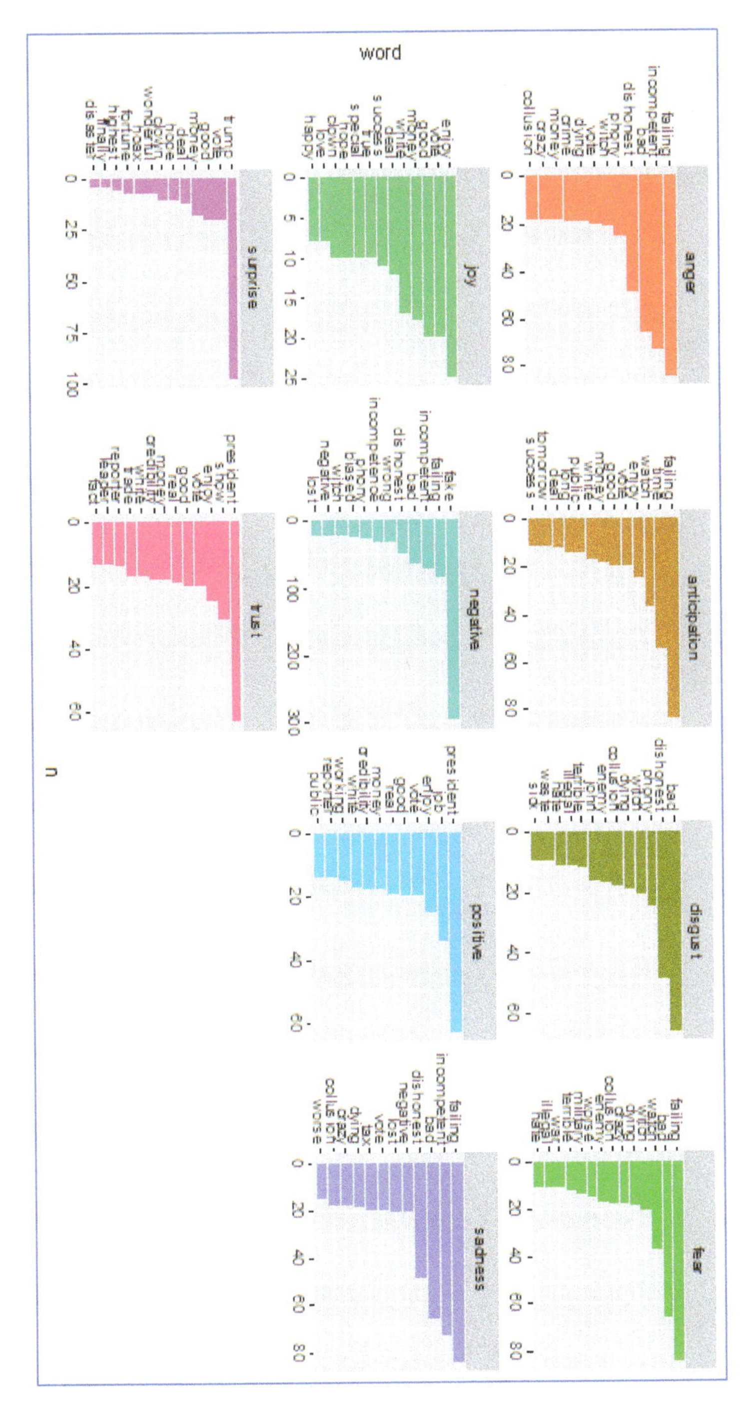

Figure 5-1. *Sentiments by word frequency*

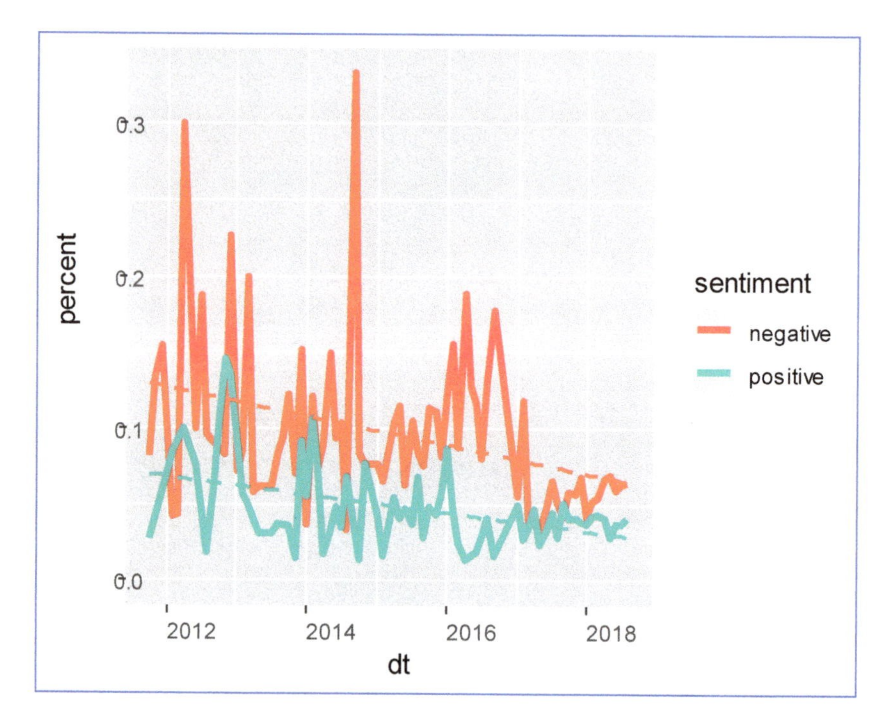

Figure 5-2. *Positive and negative sentiment over time.*

Next, we plot positive and negative words with separate histograms, using the `filter()` function with `sentiment=='positive'`. Figure 5-3 shows positive sentiment words and Figure 5-4 shows negative sentiment words.

```
pos_neg<-trump_sentiment %>%count(word, sentiment, sort=TRUE)
pos_neg %>% filter(sentiment=='positive') %>% head(20) %>%
ggplot(aes(x=word, y=n)) + geom_bar(stat="identity",
        fill="green4") +
        theme(axis.text.x=element_text(angle=90)) +
        labs(title="Most occuring Positive Words", y="count")
```

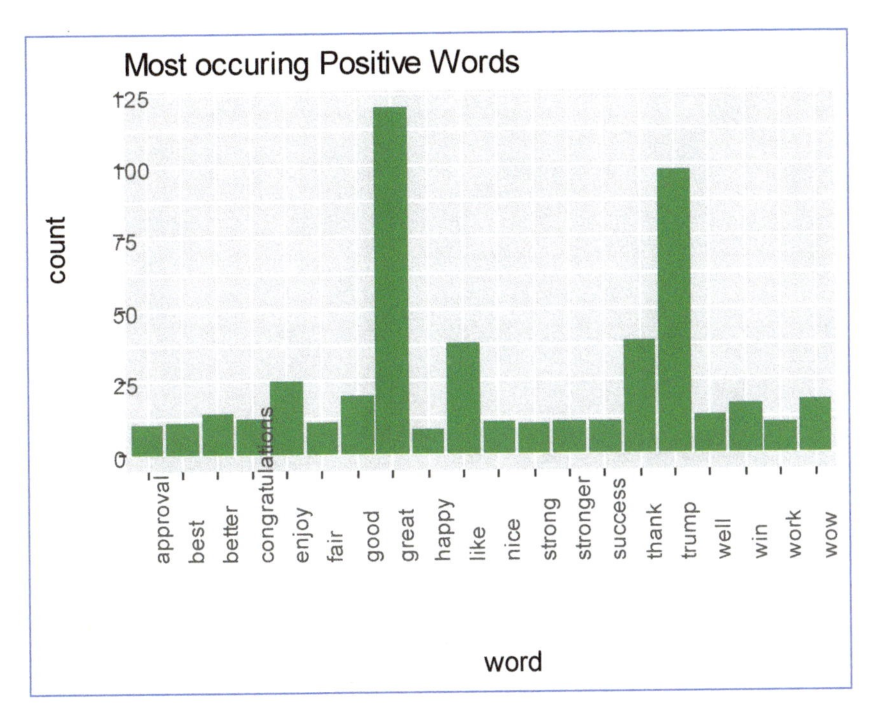

Figure 5-3. *Positive words with frequencies great than 10.*

We generate the histogram of negative words by changing the `filter()` function to `sentiment=='negative'`. Notice that among the negative words are "fake", "false", and "phony", which could be viewed a synonym. With this knowledge, we could rerun the model and setting "phony" and "fake" as equivalent to "false."

```
pos_neg %>% filter(sentiment=='negative') %>% head(20) %>%
ggplot(aes(x=word, y=n)) +
        geom_bar(stat="identity",fill="red4") +
        theme(axis.text.x=element_text(angle=90)) +
        labs(title="Most occuring Negative Words", y="count")
```

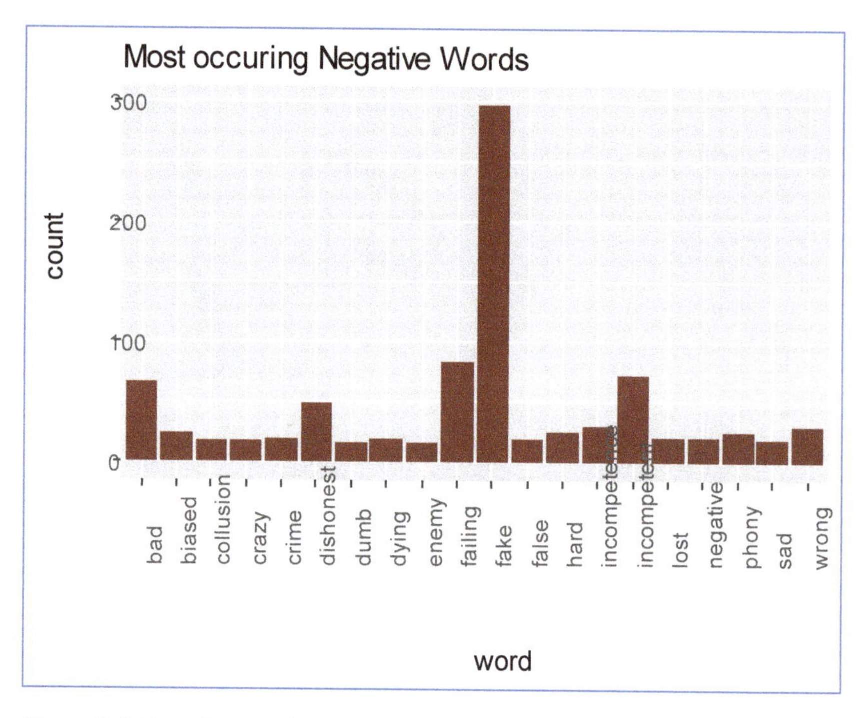

Figure 5-4. *Negative words with frequencies great than 10.*

5.5.3 Most common positive and negative words

One advantage of having the data frame with both sentiment and word is that we can analyze word counts that contribute to each sentiment. By implementing `count()` here with arguments of both word and sentiment, we find out how much each word contributed to each sentiment.

```
bing_word_counts <- trump_tweets %>%
  inner_join(get_sentiments("bing")) %>%
  count(word, sentiment, sort = TRUE) %>%
  ungroup()
```

```
Joining, by = "word"
```

```
bing_word_counts
```

```
# A tibble: 543 x 3
   word      sentiment     n
   <chr>     <chr>     <int>
 1 fake      negative    298
 2 great     positive    120
 3 trump     positive     98
 4 failing   negative     84
```

```
 5 incompetent  negative     73
 6 bad          negative     66
 7 dishonest    negative     49
 8 thank        positive     39
 9 like         positive     38
# ... with 564 more rows
```

This can be shown visually, and we can pipe straight into *ggplot2*, if we like, because of the way we are consistently using tools built for handling tidy data frames. Figure 5-5 captures these positive and negative plots.

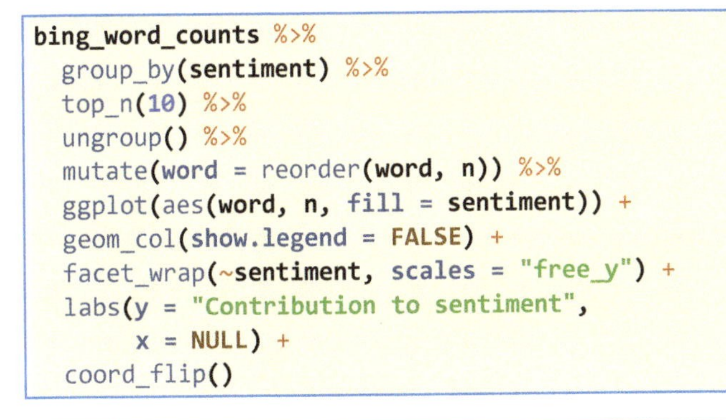

```
bing_word_counts %>%
  group_by(sentiment) %>%
  top_n(10) %>%
  ungroup() %>%
  mutate(word = reorder(word, n)) %>%
  ggplot(aes(word, n, fill = sentiment)) +
  geom_col(show.legend = FALSE) +
  facet_wrap(~sentiment, scales = "free_y") +
  labs(y = "Contribution to sentiment",
       x = NULL) +
  coord_flip()
```

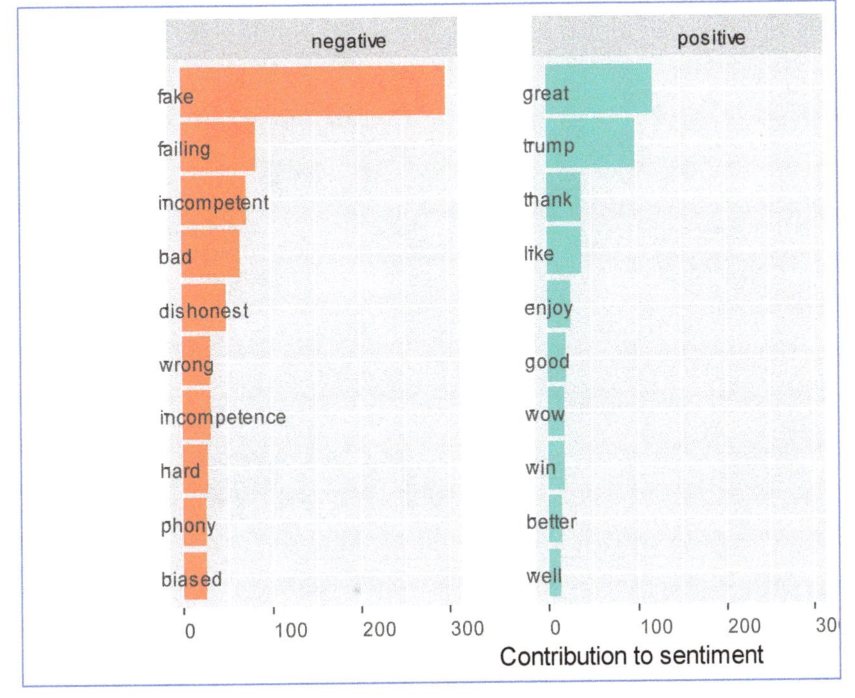

***Figure 5-5.** Top ten positive and negaitive sentiment words.*

5.5.4 Wordclouds

We've seen that this tidy text mining approach works well with ***ggplot2*** but having our data in a tidy format is useful for other plots as well. For example, consider the wordcloud package, which uses base R graphics. Let's look at the most common words in `trump_tweets`, but this time as a wordcloud, shown in Figure 5-6.

```
library(wordcloud)
trump_tweets %>%
  anti_join(stop_words) %>%
  count(word) %>%
  with(wordcloud(word, n, max.words = 100))
```

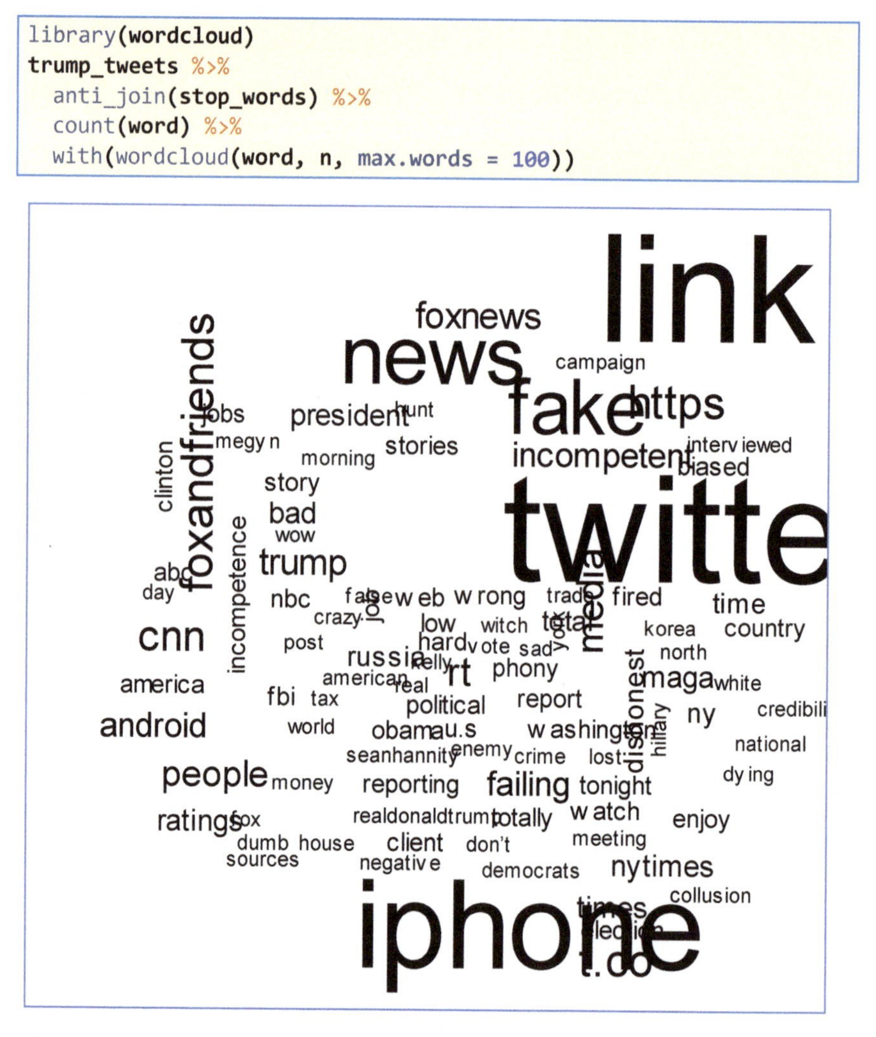

Figure 5-6. *Trump Tweet word clould.*

In other functions, such as `comparison.cloud()`, you may need to turn the data frame into a matrix with ***reshape2*** function `acast()`. Let's do the sentiment analysis to tag positive and negative words using an inner join, then find the most common positive and negative words. Until the step where we need to send the data to `comparison.cloud()`, this can all be done with *joins*, *piping*, and *dplyr* because our data is in tidy format, as shone below and pictured in Figure 5-7.

```
library(reshape2)

 Attaching package: 'reshape2'
 The following object is masked from 'package:tidyr':

     smiths
trump_tweets %>%
  inner_join(get_sentiments("bing")) %>%
  count(word, sentiment, sort = TRUE) %>%
  acast(word ~ sentiment, value.var = "n", fill = 0) %>%
  comparison.cloud(colors = c("red", "blue"),
                   max.words = 100)
```

```
 Joining, by = "word"
```

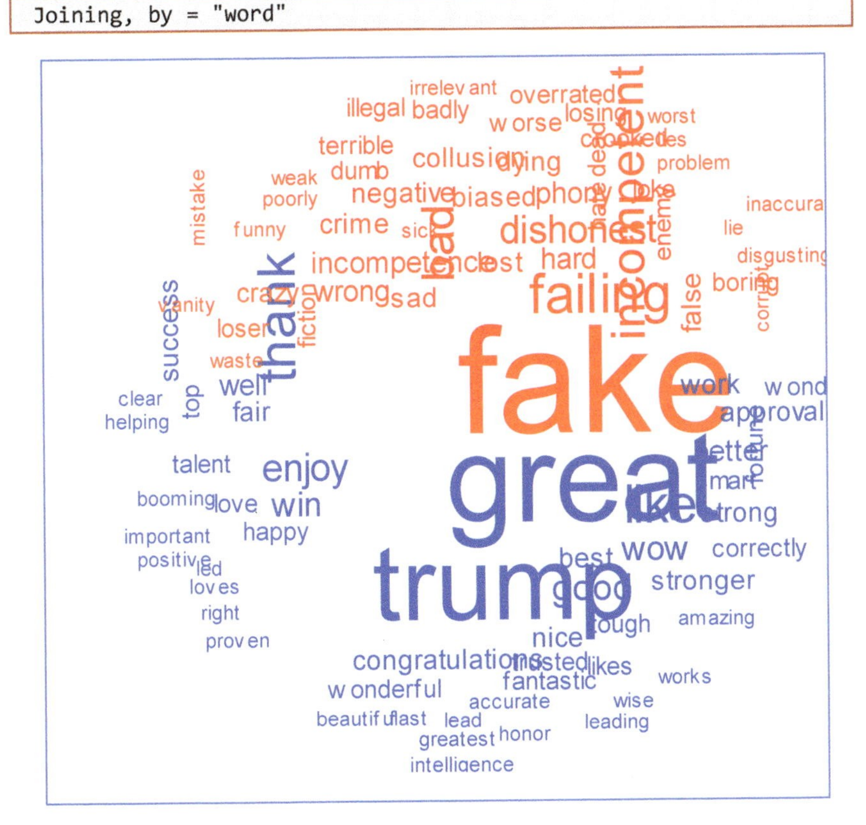

Figure 5-7. Trump Tweets comparison word cloud.

5.5.5 Word Frequencies

In this next section, we want to look at trump_tweets from the perspective of Mr. Trump's different roles, assuming that might be some sentiment differences between Candidate Trump and President Trump. The column "role" contains a flag to delineate Mr. Trump's different roles. After January 21, 2017, his role is "president" and prior to that, his role is "candidate." The next code chunk removes user-made and standard English stopwords.

```
library(tm)
library(tidyr)
my_stops <- c("https", "a", "rt", "t.co", "for", "they", stopwor
ds("en"))
trump_tweets <- trump_tweets %>%
  filter(!word %in% stop_words$word,
         !word %in% my_stops,
         !word %in% str_remove_all(stop_words$word, "'"))
```

Now we can calculate word frequencies for each Donald Trump role. First, we group by person and count how many times each role used each word. Then we use `left_join()` to add a column of the total number of words used by each role.

```
frequency <- trump_tweets %>%
  group_by(role) %>%
  count(word, sort = TRUE) %>%
  left_join(trump_tweets %>%
              group_by(role) %>%
              summarise(total = n())) %>%
  mutate(freq = n/total)
```

```
 Joining, by = "role"
frequency
 # A tibble: 3,719 x 5
 # Groups:   role [2]
    role      word              n total   freq
    <chr>     <chr>         <int> <int>  <dbl>
  1 president twitter         584  9947 0.0587
  2 president link            575  9947 0.0578
  3 president iphone          541  9947 0.0544
  4 president news            320  9947 0.0322
  5 president fake            298  9947 0.0300
  6 president foxandfriends   143  9947 0.0144
  7 president media           109  9947 0.0110
  8 president cnn             106  9947 0.0107
```

```
 9 candidate link           91  3178 0.0286
10 president foxnews        89  9947 0.00895
# ... with 3,709 more rows
```

This is a nice and tidy data frame, but we would actually like to plot those frequencies on the *x*- and *y*-axes of a plot, so we will need to use `spread()` from *tidyr* make a differently shaped data frame.

```
if(!require(tidyr)) install.packages("tidyr")
library(tidyr)
frequency <- frequency %>%
  select(role, word, freq) %>%
  spread(role, freq) %>%
  arrange(president, candidate)
frequency
```

```
# A tibble: 3,210 x 3
   word      candidate president
   <chr>         <dbl>     <dbl>
 1 18         0.000315  0.000101
 2 admitted   0.000315  0.000101
 3 agreed     0.000315  0.000101
 4 allowing   0.000315  0.000101
 5 articles   0.000315  0.000101
 6 attempt    0.000315  0.000101
 7 average    0.000315  0.000101
 8 basic      0.000315  0.000101
 9 bob        0.000315  0.000101
10 buy        0.000315  0.000101
# ... with 3,200 more rows
```

Now this is ready for us to plot. We will use `geom_jitter()` so that we don't see the discreteness at the low end of frequency as much, and `check_overlap = TRUE`, so the text labels don't all print out on top of each other (only some will print). Using *scales*, this plot is shown in Figure 5-8.

```
if(!require(scales)) install.packages("scales")
library(scales)
ggplot(frequency, aes(president, candidate)) +
  geom_jitter(alpha = 0.1, size = 2.5, width = 0.25, height = 0.
25) +
  geom_text(aes(label = word), check_overlap = TRUE, vjust = 1.5
) +
  scale_x_log10(labels = percent_format()) +
  scale_y_log10(labels = percent_format()) +
  geom_abline(color = "red")
```

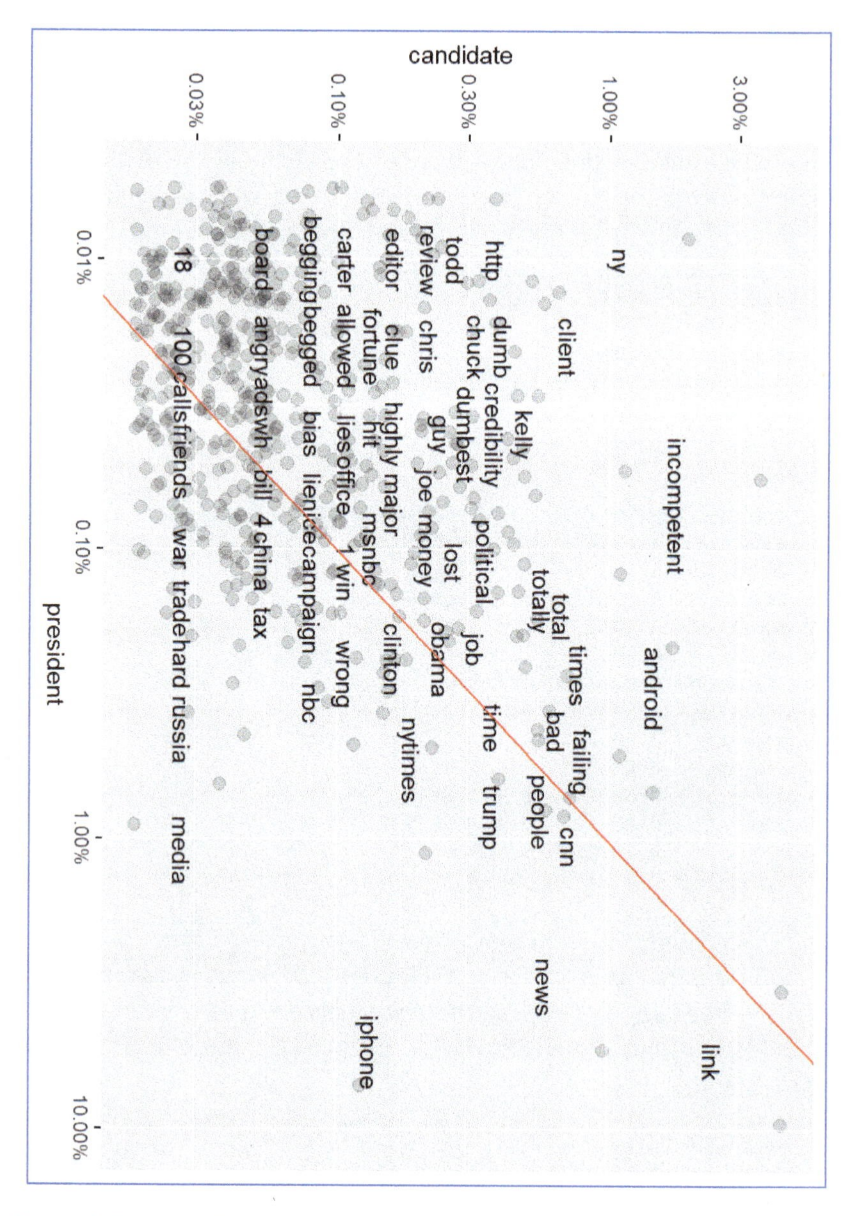

Figure 5-8. *Trump Tweets words by candidacy versus presidency.*

Words near the line are used with about equal frequencies by President Trump and Candidate Trump, while words far away from the line are used much more by one person compared to the other. Words, hashtags, and usernames that appear in this plot are ones that we have both used at least once in tweets.

5.5.6 Comparing word usage

We just made a plot comparing raw word frequencies over our whole Twitter histories; now let's find which words are more or less likely to come from each Mr. Trump's roles, using the log odds ratio. Here, we count how many times each role uses each word and keep only the words used more than 10 times. After a `spread()` operation, we can calculate the log odds ratio for each word, using log odd ratio =

$$\ln\left(\frac{\left(n+\frac{1}{total+1}\right)president}{\left(\frac{n+1}{total+1}\right)candidate}\right)$$

where n is the number of times the word in question is used by each role and the total indicates the total words for each role.

```
my_stops <- c("https", "http", "a", "rt", "t.co", "for", "they",
"north", "ny", stopwords("en"))
trump_tweets <- trump_tweets %>%
  filter(!word %in% stop_words$word,
         !word %in% my_stops,
         !word %in% str_remove_all(stop_words$word, "'"))
word_ratios <- trump_tweets %>%
  filter(!str_detect(word, "^@")) %>%
  count(word, role) %>%
  group_by(word) %>%
  filter(sum(n) >= 10) %>%
  ungroup() %>%
  spread(role, n, fill = 0) %>%
  mutate_if(is.numeric, funs((. + 1) / (sum(.) + 1))) %>%
  mutate(logratio = log(candidate / president)) %>%
  arrange(desc(logratio))
word_ratios %>%
  arrange(abs(logratio))
```

```
# A tibble: 203 x 4
   word     candidate president logratio
   <chr>        <dbl>     <dbl>    <dbl>
 1 cnn        0.0191    0.0192  -0.00422
 2 people     0.0150    0.0145   0.0330
 3 polls      0.00205   0.00197  0.0371
 4 china      0.00205   0.00215 -0.0499
 5 coverage   0.00205   0.00215 -0.0499
 6 million    0.00205   0.00215 -0.0499
 7 campaign   0.00273   0.00287 -0.0499
 8 world      0.00273   0.00287 -0.0499
# ... with 195 more rows
```

What are some words that have been about equally likely to come from Candidate Trump or President Trump? Using, the log odds ratio, we show this in Figure 5-9.

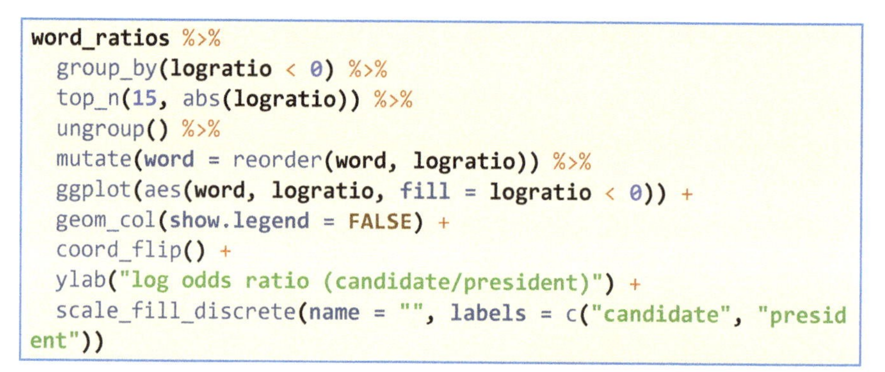

```
word_ratios %>%
  group_by(logratio < 0) %>%
  top_n(15, abs(logratio)) %>%
  ungroup() %>%
  mutate(word = reorder(word, logratio)) %>%
  ggplot(aes(word, logratio, fill = logratio < 0)) +
  geom_col(show.legend = FALSE) +
  coord_flip() +
  ylab("log odds ratio (candidate/president)") +
  scale_fill_discrete(name = "", labels = c("candidate", "presid
ent"))
```

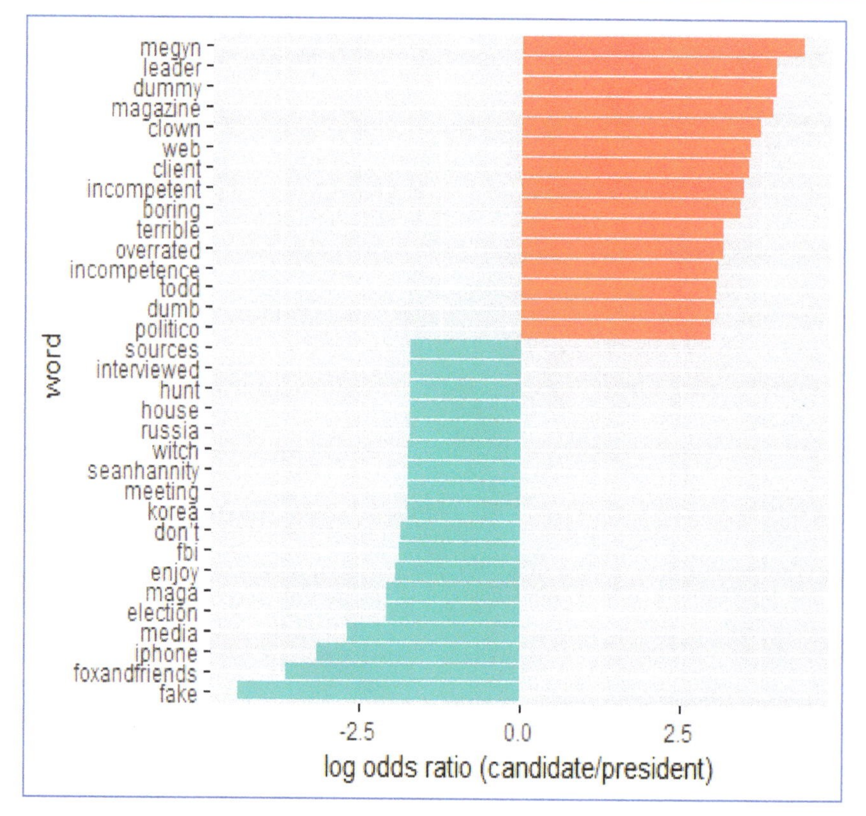

Figure 5-9. Plot of the log odds ratio for candidacy and versus presidency.

5.5.7 Changes in word use

The section above looked at overall word use, but now let's ask a different question. Which words' frequencies have changed the fastest in Mr.Trumps Twitter feeds? Or to state this another way, which words has he tweeted about at a higher or lower rate as time has passed in his different roles? To do this, we will define a new time variable in the data frame that defines which unit of time each Tweet was posted in. We can use `floor_date()` from lubridate to do this, with a unit of our choosing; using 1 month seems to work well for set of tweets.

After we have the time bins defined, we count how many times each of us used each word in each time bin. After that, we add columns to the data frame for the total number of words used in each time bin by each role and the total number of times each word was used by each role. We can then `filter()` to only keep words used at least some minimum number of times (30, in this case).

```
words_by_time <- trump_tweets %>%
  filter(!str_detect(word, "^@")) %>%
  mutate(time_floor = floor_date(date, unit = "1 month")) %>%
  count(time_floor, role, word) %>%
  group_by(role, time_floor) %>%
  mutate(time_total = sum(n)) %>%
  group_by(role, word) %>%
  mutate(word_total = sum(n)) %>%
  ungroup() %>%
  rename(count = n) %>%
  filter(word_total > 30)
words_by_time
```

```
# A tibble: 482 x 6
   time_floor role   word   count time_total word_total
   <date>     <chr>  <chr>  <int>      <int>      <int>
 1 2011-10-01 candi~ incom~     2         24         66
 2 2011-10-01 candi~ link       2         24         91
 3 2011-11-01 candi~ incom~     1          9         66
 4 2011-11-01 candi~ link       1          9         91
 5 2012-06-01 candi~ link       1         20         91
 6 2012-06-01 candi~ twitt~     1         20         86
 7 2012-08-01 candi~ incom~     1         28         66
 8 2012-08-01 candi~ link       1         28         91
 9 2012-09-01 candi~ incom~     4         49         66
```

```
10 2012-09-01 candi~ link        4          49           91
# ... with 472 more rows
```

Each row in this data frame corresponds to one role using one word in a given time bin. The count column tells us how many times that role used that word in that time bin, the `time_total` column tells us how many words that role used during that time bin, and the `word_total` column tells us how many times that person used that word over the whole year. This is the data set we can use for modeling.

We can use `nest()` from *tidyr* to make a data frame with a list column that contains little miniature data frames for each word. Let's do that now and take a look at the resulting structure.

```
nested_data <- words_by_time %>%
  nest(-word, -role)
nested_data
```

```
# A tibble: 25 x 3
   role      word        data
   <chr>     <chr>       <list>
 1 candidate incompetent <tibble [31 x 4]>
 2 candidate link        <tibble [39 x 4]>
 3 candidate twitter     <tibble [36 x 4]>
 4 candidate android     <tibble [29 x 4]>
 5 president cnn         <tibble [18 x 4]>
 6 president failing     <tibble [15 x 4]>
 7 president fake        <tibble [19 x 4]>
 8 president foxnews     <tibble [19 x 4]>
 9 president iphone      <tibble [19 x 4]>
10 president link        <tibble [20 x 4]>
# ... with 15 more rows
```

This data frame has one row for each role-word combination; the data column is a list column that contains data frames, one for each combination of role and word. Let's use `map()` from purrr (Henry and Wickham 2018) to apply our modeling procedure to each of those little data frames inside our big data frame. This is count data so let's use `glm()` with `family = "binomial"` for modeling.

```
if(!require(purrr)) install.packages("purrr")
library(purrr)
nested_models <- nested_data %>%
  mutate(models = map(data, ~ glm(cbind(count, time_total) ~ tim
e_floor, ., family = "binomial")))
```

```
nested_models
# A tibble: 25 x 4
   role      word        data              models
   <chr>     <chr>       <list>            <list>
 1 candidate incompetent <tibble [31 x 4]> <S3: glm>
 2 candidate link        <tibble [39 x 4]> <S3: glm>
 3 candidate twitter     <tibble [36 x 4]> <S3: glm>
 4 candidate android     <tibble [29 x 4]> <S3: glm>
 5 president cnn         <tibble [18 x 4]> <S3: glm>
 6 president failing     <tibble [15 x 4]> <S3: glm>
 7 president fake        <tibble [19 x 4]> <S3: glm>
 8 president foxnews     <tibble [19 x 4]> <S3: glm>
 9 president iphone      <tibble [19 x 4]> <S3: glm>
10 president link        <tibble [20 x 4]> <S3: glm>
# ... with 15 more rows
```

Now notice that we have a new column for the modeling results; it is another list column and contains glm objects. The next step is to use `map()` and `tidy()` from the ***broom*** package to pull out the slopes for each of these models and find the important ones. We are comparing many slopes here and some of them are not statistically significant, so let's apply an adjustment to the p-values for multiple comparisons.

```
if(!require(broom)) install.packages("broom")
 Loading required package: broom
library(broom)
library(tidyr)
slopes <- nested_models %>%
  unnest(map(models, tidy)) %>%
  filter(term == "time_floor") %>%
  mutate(adjusted.p.value = p.adjust(p.value))
```

Now let's find the most important slopes. Which words have changed in frequency at a moderately significant level in our tweets?

```
top_slopes <- slopes %>%
  filter(adjusted.p.value < 0.05)
top_slopess
```

role <chr>	word <chr>	term <chr>	estimate <dbl>	std.error <dbl>
candidate	incompetent	time_floor	-0.0008457277	0.0002855362
candidate	link	time_floor	-0.0008922026	0.0002349048
candidate	twitter	time_floor	-0.0009304188	0.0002706097
candidate	android	time_floor	-0.0014821153	0.0004195812
president	failing	time_floor	-0.0042022246	0.0008896979
president	iphone	time_floor	-0.0010930502	0.0003032900
president	link	time_floor	-0.0013127521	0.0002774002
president	nytimes	time_floor	-0.0058712637	0.0010400612
president	twitter	time_floor	-0.0011893092	0.0002756080
president	foxandfriends	time_floor	-0.0034668078	0.0005628461

1-10 of 10 rows | 1-5 of 8 columns

5.5.8 Evaluating Changes using Slope

To visualize our results, we can plot the word usage for both Candidate Trump and President Trump over the Tweet-period in Figure 5-10.

```
words_by_time %>%
  inner_join(top_slopes, by = c("word", "role")) %>%
  filter(role == "candidate") %>%
  ggplot(aes(time_floor, count/time_total, color = word)) +
       geom_line(size = 1.3) +
       labs(x = NULL, y = "Word frequency")
```

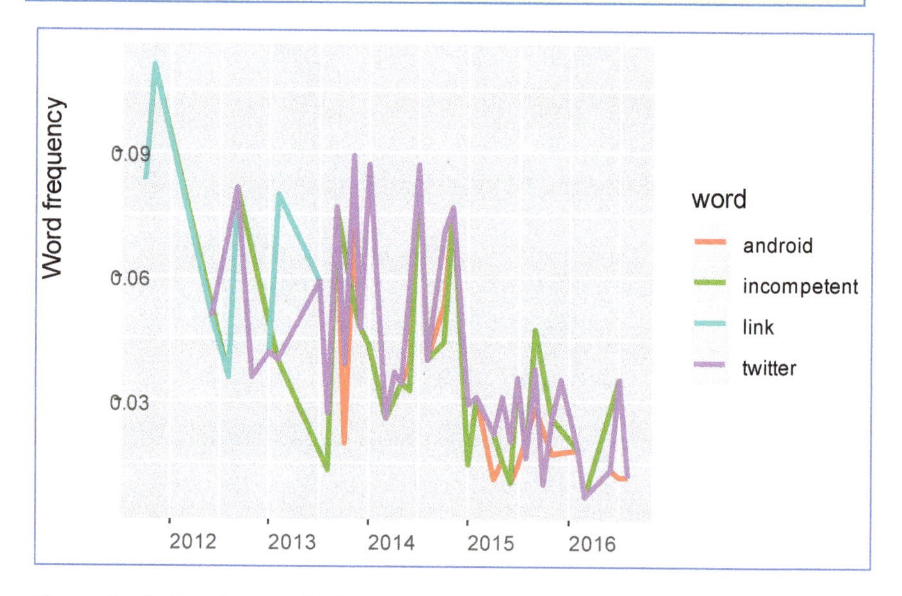

Figure 5-10. *Word usage for both Candidate Trump and President Trump over the Tweet-period.*

Now let’s plot words that have changed frequency in President Trump's tweets, which we show in **Figure 5-11**.

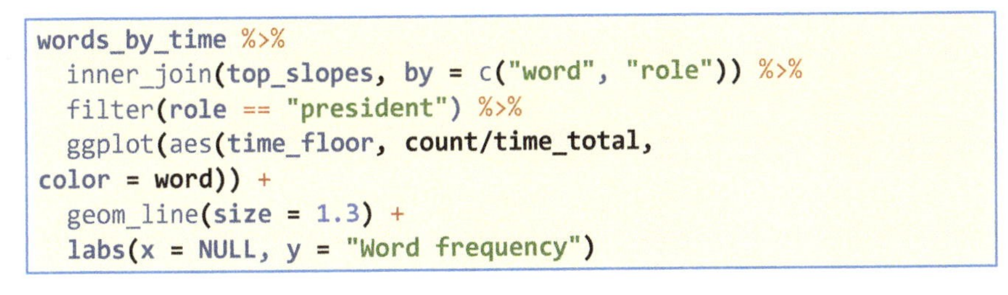

```
words_by_time %>%
  inner_join(top_slopes, by = c("word", "role")) %>%
  filter(role == "president") %>%
  ggplot(aes(time_floor, count/time_total,
color = word)) +
  geom_line(size = 1.3) +
  labs(x = NULL, y = "Word frequency")
```

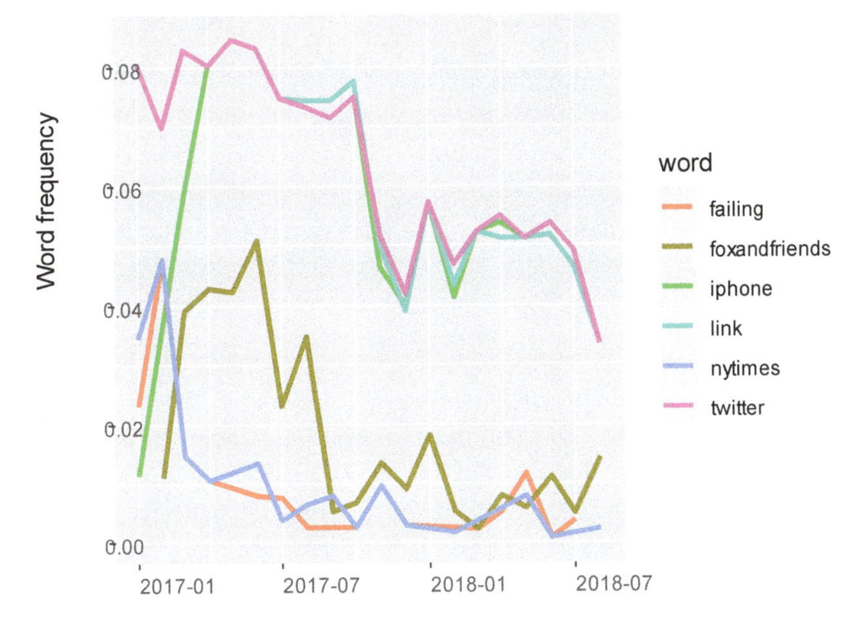

Figure 5-11. *Words that have changed frequency in President Trump's tweets.*

5.6 Example 2: IsisFanboy

So-called ISIS Fanboy have been actively tweeting all over the world. A Fanboy is a boy or man who ardently supports a single hobby, ideology, movement, etc. ISIS released TweetMovie from Twitter, a "normal" day when two ISIS operatives murdered a priest saying mass in a French church. The two attackers, who said they were from the so-called Islamic State (IS), slit Fr Jacques Hamel's throat during a morning Mass, according to French officials. (BBC, 2016)

Fanboy tweets obviously would suggest certain sentiments. However, there are also many tweets that perhaps counterpoise the ISIS Fanboy tweets. Kaggle release a selection of data from the site and made it available to Kaggle users. You can find the dataset in my GitHub repository, with the filename: ISISFanboy. This data set is intended to be a counterpoise to the *How ISIS Uses Twitter* data set. That data set contains 17k tweets alleged to originate with "100+ anti-ISIS Fanboys." This is not a perfect counterpoise as it almost surely contains a small number of pro-ISIS Fanboy tweets.

Our goal in this chapter is to perform a counterpoise analysis to pro-ISIS tweets. A counterpoise provides a balance or backdrop against which to measure a primary object, in this case the original pro-ISIS data. So, if anyone wants to discriminate between pro-ISIS tweets and other tweets concerning ISIS, we will need to model the original pro-ISIS data or signal against the counterpoise which is signal + noise.

5.6.1 Loading the Data

It is always a good idea to make sure you place data in the R working directory. This makes it easy to load data and locate your results—it keeps things organized. To check the working directory type:

```
getwd()
```

```
 [1] "C:/Users/jeff/Documents/VIT_University/IsisFanboy"
```

If you are in the directory you want to be in, you do not need to do anything. However, if you want to be in a different directory, the `setwd()` command, like:

```
setwd("C:/Users/jeff/Documents/VIT_University/IsisFanboy/")
```

Note that if you begin your analysis by setting up a project as describe on pages XX-YY. By doing this, the working directory will be set to your project directory by default.

Now we load the data using the `read.csv()` function:

```
tweets<-read.csv("AboutIsis.csv",stringsAsFactor=FALSE)
str(tweets)
```

Next, we compactly display the internal structure of or data (an R object), as a diagnostic function and an alternative to `summary()`:

```
str(tweets) # output not shown
```

Ideally, only one line for each 'basic' structure is displayed. It is especially well suited to compactly display the (abbreviated) contents of nested lists. The idea is to give reasonable output for any R object.

5.6.2 *Fixing the dates*

We now use the function `as.Date()` to convert between character representations and objects of class "Date" representing calendar dates. In other words, the dates in the CSV file are entered as character strings and we must convert them to date data, using day/month/year and hour/minute/second format.

```
tweets$date<-as.Date(tweets$time, "%d/%m/%Y %H:%M:%S")
head(tweets)
```

```
[partial output]
        date
 1 2016-11-07
 2 2016-11-07
 3 2016-11-07
 4 2016-11-07
 5 2016-11-07
 6 2016-11-07
```

5.6.3 *Tokenization*

Tokenization is the process of breaking up a sequence of strings into pieces such as words, keywords, phrases, symbols and other elements called tokens. Tokens can be individual words, phrases or even whole sentences. In the process of tokenization, some characters like punctuation marks are discarded. Consequently, it appears that we are "undoing" what we did in the previous step with the dates. This is not the case, however. Rather, we are taking the date strings, converting them to dates, breaking them into dates and times, and then creating one string for dates and another for times. This is not entirely necessary since al the dates are the same—we could have just pulled out the times. Note that the sequence of tweets could be important.

```
tweets$date<-as.Date(tweets$time, "%d/%m/%Y %H:%M:%S")
tidy_tweets<-tweets %>%group_by(name,username,tweetid)%>%
mutate(ln=row_number())%>%
unnest_tokens(word,tweets)%>%
ungroup()
head(tidy_tweets,5)
```

The result from this process yields dates like:

```
# A tibble: 5 x 7
  name      username tweetid date         time          ln     word
  <chr>     <chr>     <dbl>   <date>      <time>        <int> <chr>
1 jairoes~ jairoes~ 7520…    2016-07-12  1:30:19 AM    1       how
```

```
2 jairoes~ jairoes~ 7520…   2016-07-12  1:30:19 AM   1      isis
3 jairoes~ jairoes~ 7520…   2016-07-12  1:30:19 AM   1   related
4 jairoes~ jairoes~ 7520…   2016-07-12  1:30:19 AM   1        to
5 jairoes~ jairoes~ 7520…   2016-07-12  1:30:19 AM   1     islam
```

Now, we split the dates and times. Note that times include the date, like 7/11/2016 8:45:39 AM, else we would not know what day it is.

```
t_wrd<-tidy_tweets %>%count(word,sort=TRUE)
head(t_wrd,5)
```

From `head(t_wrd,5)`, the output shows what we intended:

```
 A tibble: 5 x 7
  word     n
  <chr>    <int>
1 isis     87813
2  the     68181
3   in     51784
4   to     38870
5   of     35842
```

Note that **tibbles** are a modern take on data frames. They keep the features that have stood the test of time and drop the features that used to be convenient but are now frustrating (i.e. converting character vectors to factors). **tidy_tweets** contains all words like ‘the’,‘is’,‘are’ etc. So, we’ll take only the sentimental words by joining with lexicons.

5.6.4 Removing common words

The dataframe, **tidy_tweets**, contains all words like 'the', 'is', 'are' etc. So, let’s get only the sentimental words by joining with lexicons. One of the readily available lexicons in R is Bing Lexicon. The Bing lexicon categorizes words in a binary fashion into positive and negative categories.

```
tweets_sentiment<-tidy_tweets%>% inner_join(get_sentiments("bing
"))
```

Some of the words in tidy_tweets include:

```
tidy_tweets$word
```

```
   [2] "influence"
   [3] "on"
   [4] "the"
```

```
  [5] "decline"
When we applied the step giving us t_wrd, the output looked like
this:
t_wrd
# A tibble: 90,942 x 2
    word        n
    <chr>  <int>
 1  isis   113117
 2  ã       93958
 3  rt      86464
 4  the     67765
 5  à       58118
```

Note that these are some of the words that will be stripped out by the inner join with the lexicon. When we apply the inner join, we get:

```
tweets_sentiment$word
```

```
   [1] "truthful"        "easy"            "promise"
   [4] "victory"         "dislike"         "attack"
   [7] "available"       "breaking"        "supports"
  [10] "best"            "drastically"     "support"
  [13] "killing"         "love"            "holy"
```

Notice that words like "on", "the", ã, rt, and "as" no longer appear.

5.6.5 Word frequencies and word clouds

Now we want to start analyzing our data. We start by examining word frequencies. A good graphical representation of word frequencies is the "word cloud." We begin by getting a count of the words that remain in our dataset.

```
tot_wrd<-tweets_sentiment%>%count(word,sort=TRUE)
```

```
 A tibble: 1,609 x 2
   word          n
   <chr>     <int>
 1 killed     1322
 2 breaking    558
 3 attack      539
 4 like        316
 5 support     236
```

Now we can generate a word cloud. Word clouds show all the words (unless you restrict the number or words) and represent the frequencies by the sizes of the words within the cloud shown in Figure 5-12. Notice the relative sized of "killed", "attack", and "death." See if you can find "peace" and "destroy."

```
wordcloud (tot_wrd$word,tot_wrd$n, min.freq =25,
    scale=c(5, .2), random.order = FALSE,
    random.color = FALSE,
    colors = brewer.pal(8, "Dark2")
)
```

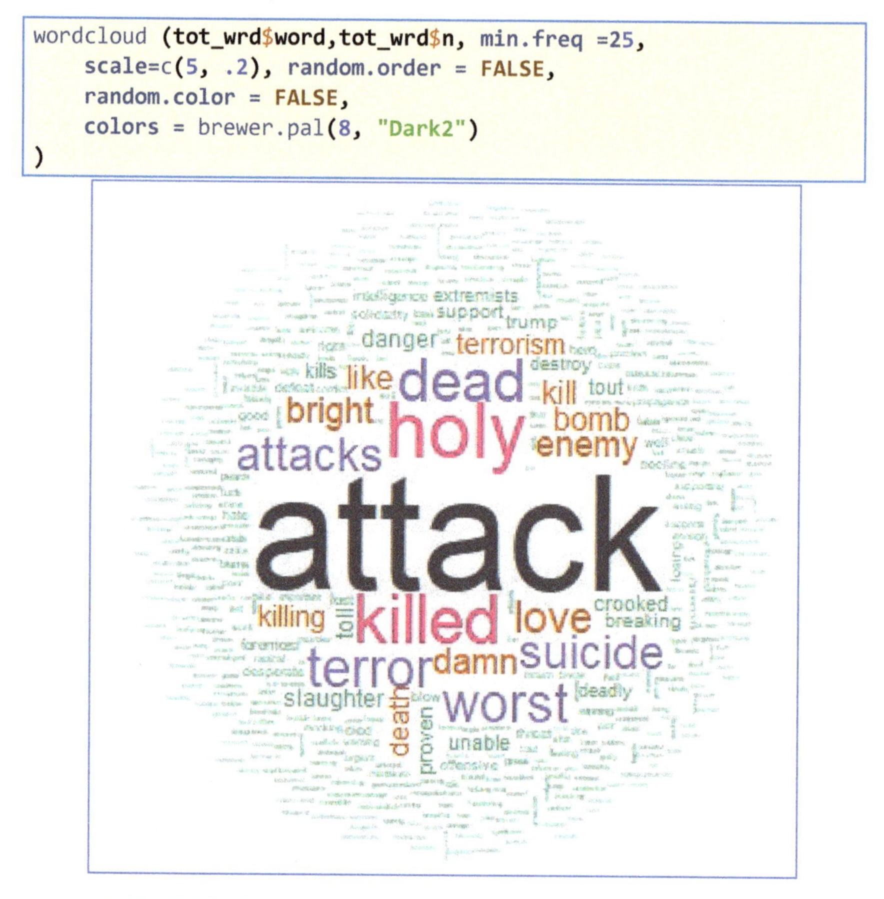

Figure 5-12. ISIS Fanboy tweets using a word cloud.

```
pos_neg<-tweets_sentiment %>%count(word,sentiment,sort=TRUE)
```

5.6.6 Plotting the most occuring positive and negative words

Next, we will construct bar charts depicting the most frequent words in the corpus of ISIS related text messages. Here we use `ggplot`, which allows us to aesthetics, specify plt type, designate plot theme, and designate labels. For instance, we use `labs()` to apply the label "Most occurring Positive Words." Also, we specify the number of words to include on the chart using the `head(n)` function, where n = 15. Figure

5-13 shows the positive sentiment word frequencies while Figure 5-14 shows the negative word frequencies.

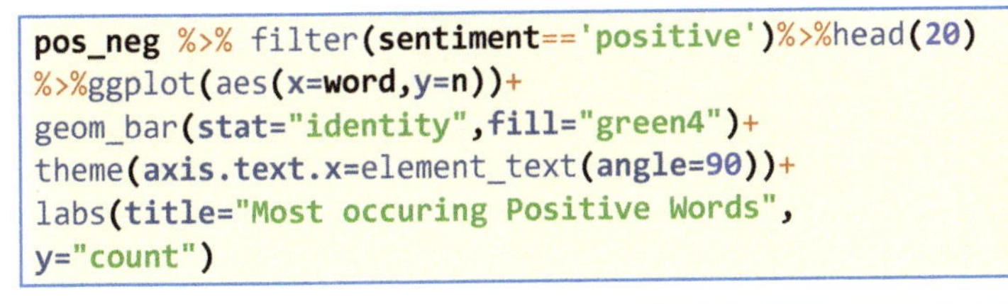

```
pos_neg %>% filter(sentiment=='positive')%>%head(20)
%>%ggplot(aes(x=word,y=n))+
geom_bar(stat="identity",fill="green4")+
theme(axis.text.x=element_text(angle=90))+
labs(title="Most occuring Positive Words",
y="count")
```

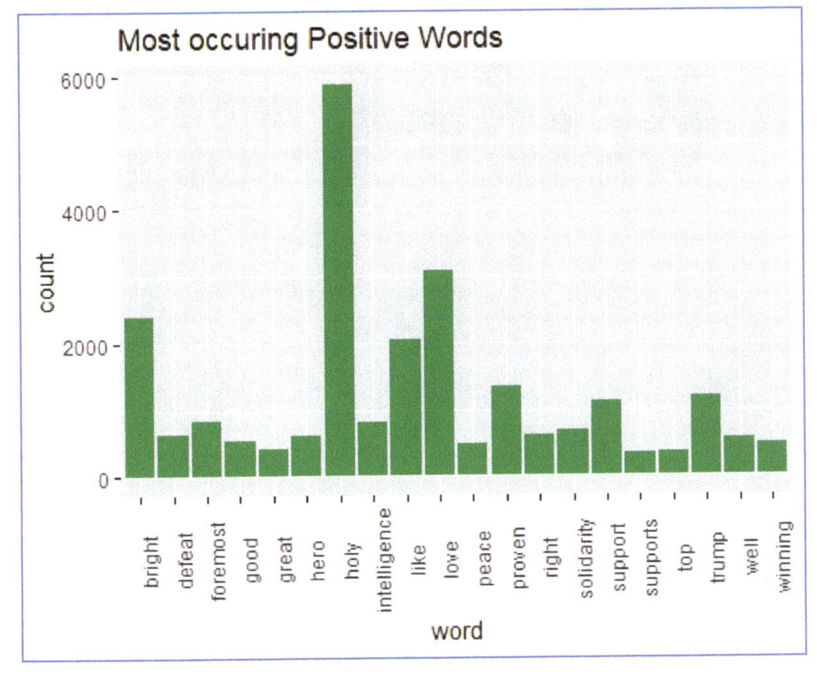

__Figure 5-13.__ ISIS Fanboy positive sentiment word frequencies.

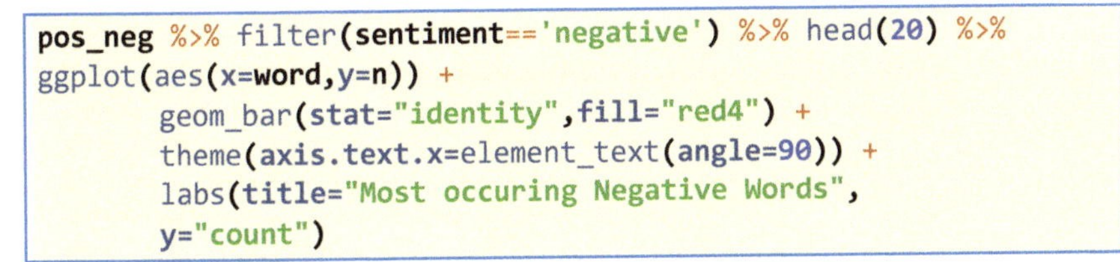

```
pos_neg %>% filter(sentiment=='negative') %>% head(20) %>%
ggplot(aes(x=word,y=n)) +
        geom_bar(stat="identity",fill="red4") +
        theme(axis.text.x=element_text(angle=90)) +
        labs(title="Most occuring Negative Words",
        y="count")
```

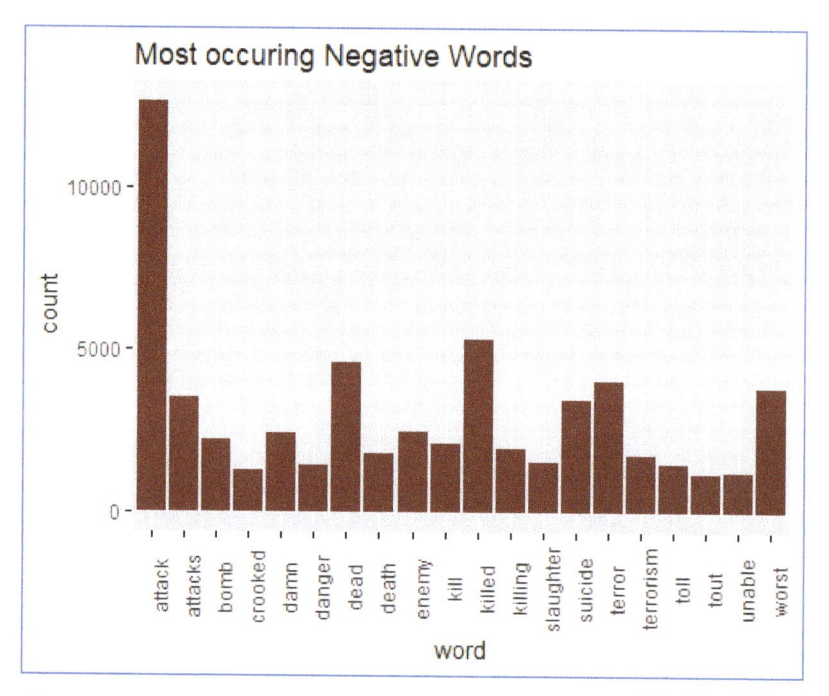

Figure 5-14. *ISIS Fanboy negative sentiment word frequencies.*

5.6.7 New Bigram Counts

Our next step is to get bigrams. We can examine the most common bigrams using *dplyr* function `count()`. As we might expect, a lot of the most common bigrams are pairs of common (uninteresting) words, such as "of the" and "to be" (stop-words, see **Definition 2.8**). This is a useful time to use *tidyr* function `separate()`, which splits a column into multiple based on a delimiter. This lets us separate it into two columns, "word1" and "word2", at which point we can remove cases where either is a stop-word. We will use the *stringr* function `str_replace_all()` to remove all non-alphanumeric characters, like à, ã, ø, ù, å, etc.

```
# Unnest tokens
demo_bigrams <- unnest_tokens(tweets, input = tweets,
       output = bigram, token = "ngrams", n=2)

demo_bigrams %>%
  count(bigram, sort = TRUE)

bigrams_separated <- demo_bigrams %>%
  separate(bigram, c("word1", "word2"), sep = " ")
```

```
# Remove stopwords
bigrams_filtered <- bigrams_separated %>%
  filter(!word1 %in% stop_words$word) %>%
  filter(!word2 %in% stop_words$word)

# Remove special characters
library(stringr)
bigrams_filtered <- str_replace_all(bigrams_filtered,
       "[^[:alnum:]]", " ")

# new bigram counts:
bigram_counts <- bigrams_filtered %>%
  count(word1, word2, sort = TRUE)
bigram_counts
```

```
# A tibble: 206,674 x 3
   word1 word2     n
   <chr> <chr> <int>
 1 à     à     49079
 2 ã     ã     43140
 3 2016  07    10569
 4 ø     ø      8012
 5 ã     ãƒ     7980
 6 ø     ù      7240
 7 https t.co   6973
 8 ù     ø      6658
 9 ãƒ    ã      5950
10 å     ã      5617
 # ... with 206,664 more rows
```

5.6.8 *Word clouds on positive and negative sentiment*

We constructed work clouds for positive and negative sentiment as we did for pro ISIS words. Figure 5-15 presents them for side by side and used the `par()` function to format them, with two columns and one row.

```
par(mfrow=c(1,2))
pos<-pos_neg %>% filter(sentiment=='positive')
neg<-pos_neg %>% filter(sentiment=='negative')
wordcloud (pos$word, pos$n, min.freq =15, scale=c(5, .2),
random.order = FALSE, random.color = FALSE,
colors = brewer.pal(9,"Spectral"))
wordcloud (neg$word,neg$n, min.freq =30, scale=c(5, .2),
       random.order = FALSE, random.color = FALSE,
       colors = brewer.pal(9, "Spectral"))
```

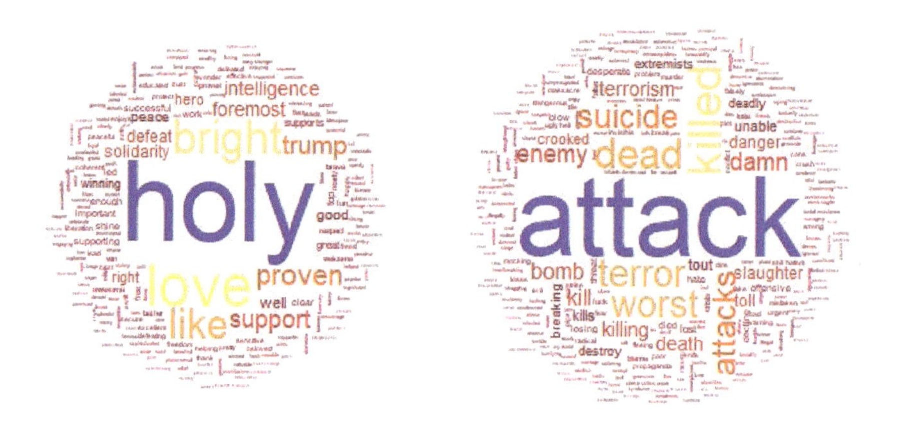

Figure 5-15. *Separate ISIS Fanboy sentiment word clouds*

5.6.9 Top 10 words contributing to different sentiments

The code below produces a plot of ten sentiments and the top five words contributing to them. We are show six sentiments in plot that follows, including "constraining," "litigious," "positive," "negative," "superfluous," and "uncertainty." From a pure sentiment perspective, this does not seem too threatening. This is coded below and Figure 5-16 shows the top six (due to readability). We get a word count by sentiment using `count()`, then we group the words by sentiment using `group_by()` and then ungroup them for plotting. We use the `mutate()` function to reorder the words

```
tidy_tweets %>% inner_join(get_sentiments("loughran")) %>%
count(word, sentiment) %>%
group_by(sentiment) %>% top_n(10) %>%
ungroup() %>% mutate(word=reorder(word,n)) %>%
ggplot(aes(x=word,y=n,fill=sentiment)) +
        geom_col(show.legend = FALSE) +
        facet_wrap(~ sentiment, scales = "free") +
        coord_flip()
```

```
Joining, by = "word"
```

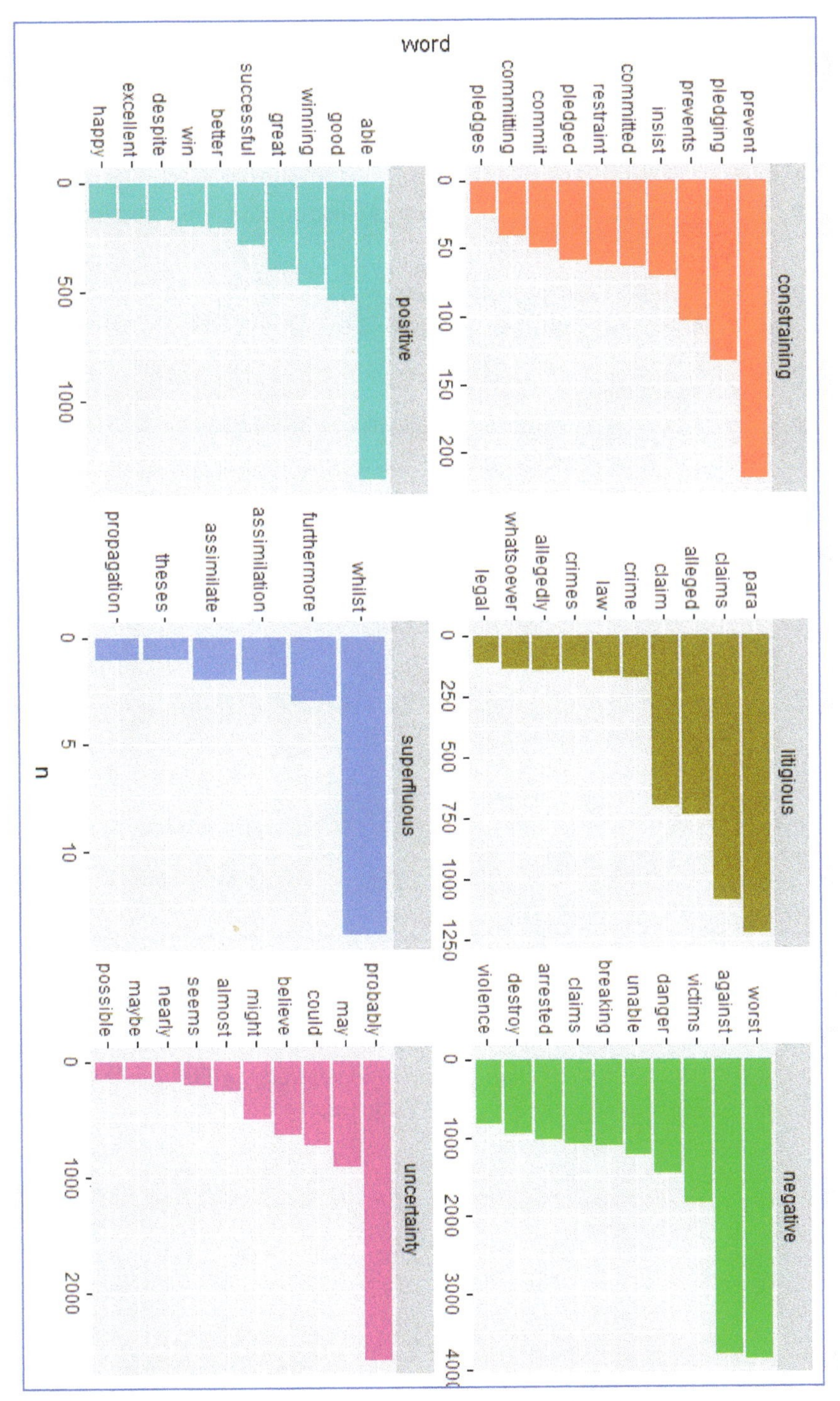

Figure 5-16. *ISIS Fanboy histograms showibg selected word frequencies.*

In this section we will look at the trend of positive and negative words over time on an expanded (by date) dataset, using the `Bing` lexicon. The `Bing` lexicon categorizes words in a binary fashion ("yes"/"no") into categories of positive and negative. The `mutate()` function adds new variables and preserves existing; `transmute()` drops existing variables. We will use `mutate()` several times throughout this section, and *ggplot* to present the results.

```
sentiment_by_time <- tidy_tweets %>%
  mutate(dt = floor_date(date, unit = "month")) %>%
  group_by(dt) %>%
  mutate(total_words = n()) %>%
  ungroup() %>%
  inner_join(get_sentiments("bing"))
```

```
Joining, by = "word"
```

The `geom_smooth()` function attempts to fit a linear model (`lm`) to both positive and negative sentiment.

```
sentiment_by_time %>%
  filter(sentiment %in% c('positive','negative')) %>%
  count(dt,sentiment,total_words) %>%
  ungroup() %>%
  mutate(percent = n / total_words) %>%
  ggplot(aes(x=dt,y=percent,col=sentiment,group=sentiment)) +
  geom_line(size = 1.5) +
  geom_smooth(method = "lm", se = FALSE, lty = 2) +
  expand_limits(y = 0)
```

Figure 5-17 shows that the negative sentiment slightly decreases over time as a linear model, but it is not fit well and essentially remains constant after an initial drop in April. The positive sentiment is fit well by a linear model and slightly increases over time.

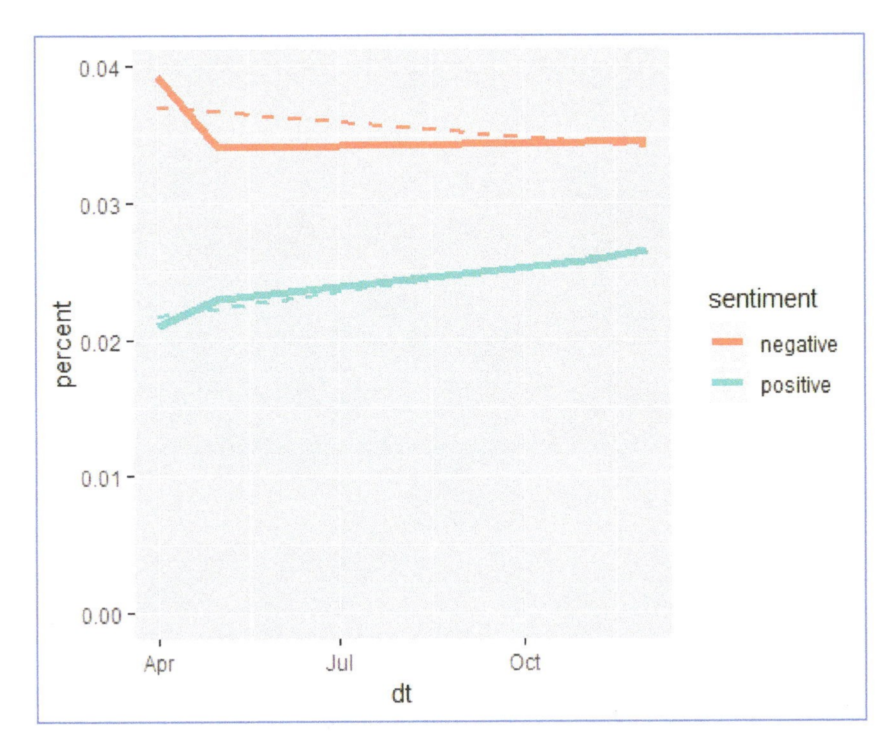

Figure 5-17. *Positive and negative sentiment over time.*

5.6.10 Sentiment proportion analysis

Percentages of positive and negative words in the tweets are generated by the code below. The pie chart in Figure 5-18 shows the percentages of positive sentiment and negative sentiment, with negative sentiment "winning" by a long shot. We use the `pie3D()` function from the ***plotrix*** package to create the pie chart.

```
library(plotrix)
perc<-tweets_sentiment %>%
count(sentiment) %>%
mutate(total=sum(n)) %>%
group_by(sentiment) %>%
mutate(percent=round(n/total,2)*100) %>% ungroup()
label <- c( paste(perc$percent[1],'%','-',
        perc$sentiment[1],sep=''),
        paste(perc$percent[2],'%','-',
        perc$sentiment[2],sep=''))
pie3D(perc$percent,labels=label,labelcex=1.1, explode=0.1)
```

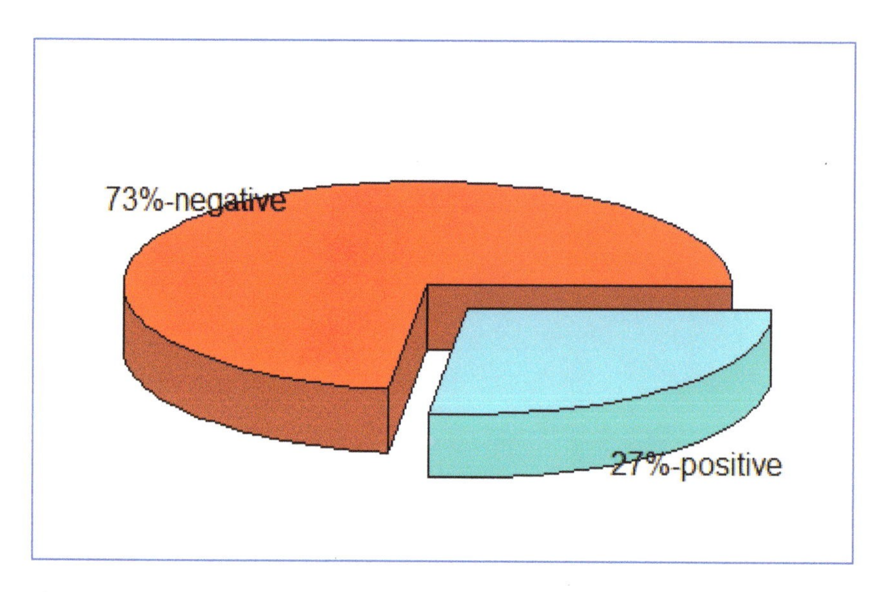

Figure 5-18. *Percentages of positive sentiment and negative sentiment.*

5.6.11 Get AFINN Setiments

```
AFINN <- get_sentiments("afinn")
not_words <- bigrams_separated %>%
  filter(word1 == "holy") %>%
  inner_join(AFINN, by = c(word2 = "word")) %>%
  count(word2, value, sort = TRUE) %>%
  ungroup()
```

5.6.12 N-grams

Recall that an n-gram is a sequence of *n* "words" taken, in order, from a body of text. If we set *n* = 2, then we get a bigram. Using `unnest_tokens()`, we now extract bigrams from the tweets and count their occurrances.

```
demo_bigrams <- unnest_tokens(tweets, input = tweets, output = b
igram, token = "ngrams", n=2)
demo_bigrams %>%
  count(bigram, sort = TRUE)
```

```
# A tibble: 348,306 x 2
  bigram             n
  <chr>          <int>
1 islamic state  10742
2 2016 07        10567
3 isis is         9484
4 during the      6530
```

```
 5 a a          6056
 6 in the       5407
 7 of ramadan   5000
 8 on the       4945
 9 holy month   4941
10 is the       4925
# ... with 348,296 more rows
```

Looking at the output, it should be clear that there are stop words and numbers in the text that form bigrams, like "a a". So, we will use the `filter()` function to filter out the stop wrods and get the new bigrams.

```
bigrams_separated <- demo_bigrams %>%
  separate(bigram, c("word1", "word2"), sep = " ")
bigrams_filtered <- bigrams_separated %>%
  filter(!word1 %in% stop_words$word) %>%
  filter(!word2 %in% stop_words$word)
bigram_counts <- bigrams_filtered %>%
  count(word1, word2, sort = TRUE)
```

```
bigram_counts
# A tibble: 210,170 x 3
   word1   word2             n
   <chr>   <chr>         <int>
 1 islamic state         10742
 2 holy    month          4941
 3 right   donaldtrump    3909
 4 isis    muslim         3645
 5 muslim  holy           3614
 6 muslim  people         3580
 7 bombing muslim         3573
 8 muslim  communities    3549
 9 dude    isis           3535
10 right     xeni            3439
# ... with 210,160 more rows
```

We can also get bigrams with specific words, like "attack", as the first of second word.

```
bigrams_separated %>%
  filter(word1 == "attack") %>%
  count(word1, word2, sort = TRUE)
```

```
# A tibble: 517 x 3
  word1  word2         n
  <chr>  <chr>     <int>
 1 attack in        4077
 2 attack slaughter 1440
 3 attack the       1046
 4 attack weeks      616
 5 attack on         417
```

```
 6 attack isis        266
 7 attack is          200
 8 attack by          196
 9 attack at          156
10 attack malaysia    155
# ... with 507 more rows
```

5.6.13 Get Sentiment Scores

The AFINN lexicon assigns words with a score that runs between -5 and 5, with negative scores indicating negative sentiment and positive scores indicating positive sentiment. Here we are looking for words preceded by "holy," as shown in Figure 5-19. We also use `ggplot()` to show the words following "holy war." We use AFINN to score the word pairs. A high negative score, like -166, would express a very strong negative sentiment, while a strong positive sentiment score might be 600. The score for "holy"-word pairs are not high in the negative direction and are 0 in the positive direction. You can see the results in Figure 5-20.

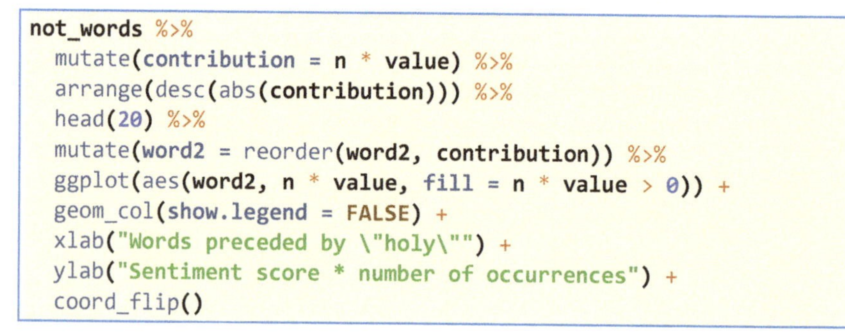

```
not_words %>%
  mutate(contribution = n * value) %>%
  arrange(desc(abs(contribution))) %>%
  head(20) %>%
  mutate(word2 = reorder(word2, contribution)) %>%
  ggplot(aes(word2, n * value, fill = n * value > 0)) +
  geom_col(show.legend = FALSE) +
  xlab("Words preceded by \"holy\"") +
  ylab("Sentiment score * number of occurrences") +
  coord_flip()
```

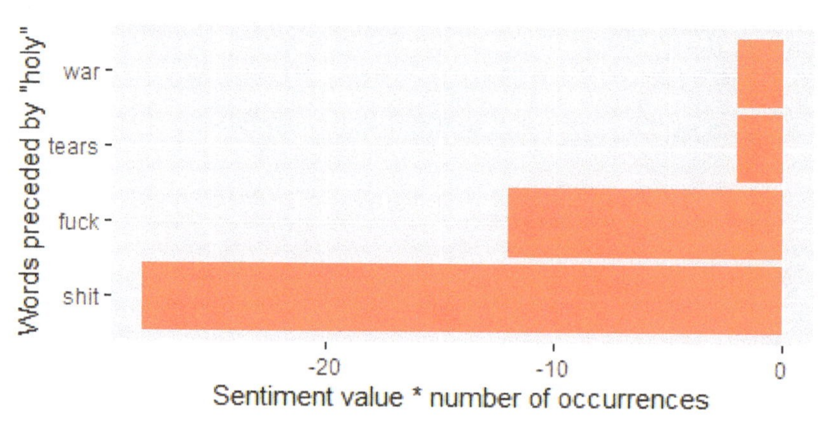

Figure 5-19. *Frequencies of words preceded by "holy."*

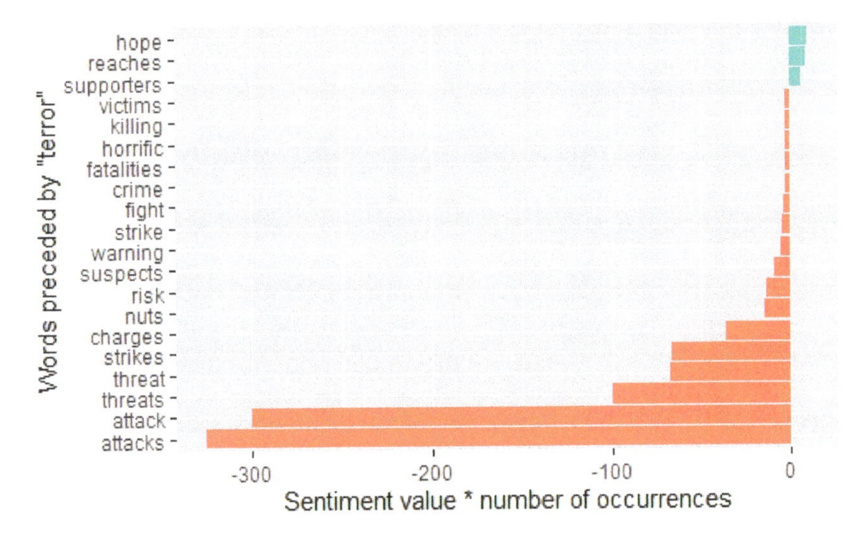

Figure 5-20. *AFINN to score the word pairs.*

5.6.14 Network of bigrams

Next, we plot the network of bigrams as shown in Figure 5-21. Figure 5-22 shows the expoded view of the red rectangle of Figure 5-21.

```
bigram_graph <- bigram_counts %>%
  filter(n > 10) %>%
  graph_from_data_frame()
```

```
IGRAPH ab56dd1 DN-- 8486 15184 --
+ attr: name (v/c), n (e/n)
+ edges from ab56dd1 (vertex names):
 [1] 2016        ->07          holy        ->month
 [3] right       ->donaldtrump isis        ->muslim
 [5] muslim      ->holy        muslim      ->people
 [7] bombing     ->muslim      muslim      ->communities
 [9] dude        ->isis        right       ->xeni
[11] xeni        ->dude        isis        ->attack
[13] 07          ->11          right       ->weteachlife
[15] weteachlife ->isis        holiest     ->month
+ ... omitted several edges
```

Now, we plot the bigrams_graph.

```
set.seed(2017)
ggraph(bigram_graph, layout = "fr") +
  geom_edge_link() +
  geom_node_point() +
  geom_node_text(aes(label = name), vjust = 0, hjust = 0)
```

The output is a little messy, but we can make out the following bigrams:

- isis attack
- isis terrorists
- isis suicide (as as a trigram: isis suicide bomber)
- muslim countries
- muslim people
- muslim bombing

These are words that occur 1000 times or more as pairs. With tweets, it is ofent difficult to remove every erroneous symbol, like é, à, ä, and ŷ, but can go back and remove them as wee see theme appear.

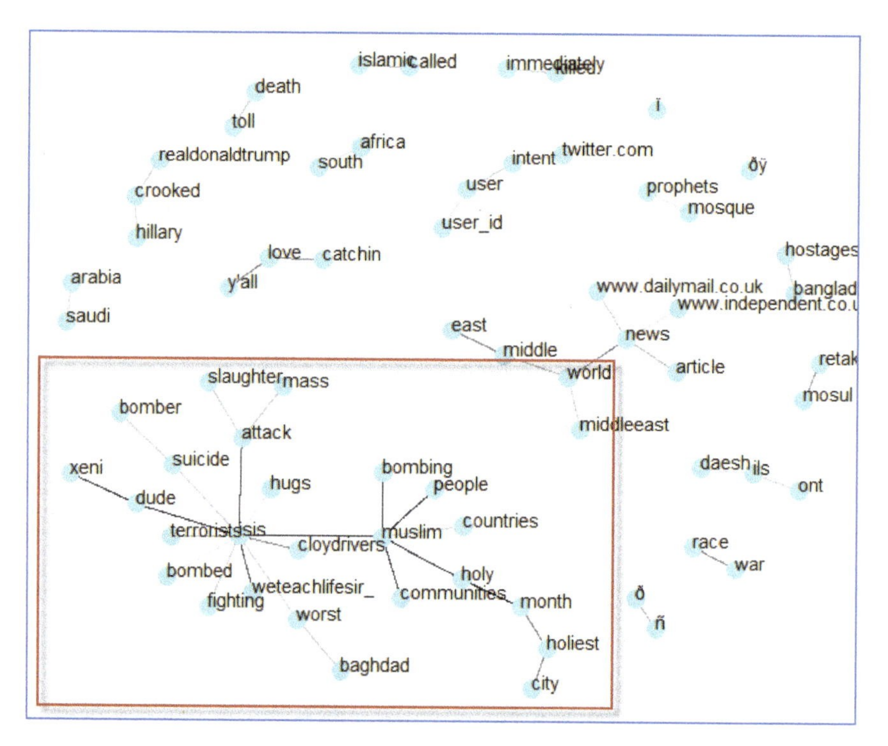

Figure 5-21. These are bigrams that occur 1000 times.

Below is the zoom-in view of the red rectangle above.

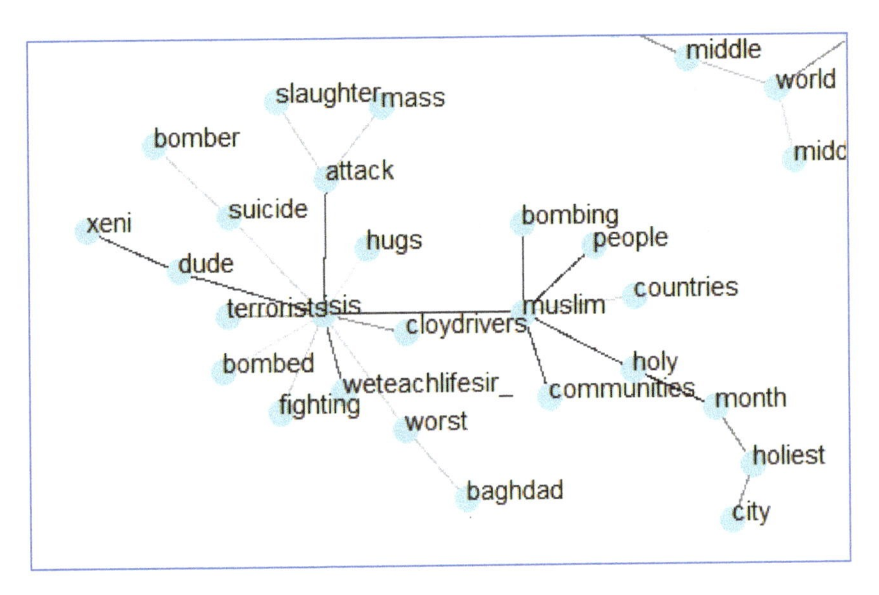

Figure 5-22. *Zoom-in view of the red rectangle in Figure 5-21*

5.7 Exercises

1. Download 250000_Plus_Tweeets from *https://github.com/stricje1/VIT_University/upload/master/Data_Analytics_2018/data* and perform Tweet Sentiment Analysis.
    a. Chart positive and negative Words over time.
    b. Chart Most common positive and negative words.
    c. Create a positive and negative Wordcloud.
    d. Compare word usage.

2. Visit this blog by Shahin Ashkiani, Bi-grams, Tri-grams, and word network, at *https://rpubs.com/Shaahin/anxiety-bigram*.

    a. Work through the example. The code for it is at the end of the blog post.

    b. Render the results as a Word document using R Markdown.

3. Rerun the Trump Tweets example for negative words setting "fake" and "phony" as synonyms of "false." Compare your results with the example results in Section 5.5.2.

Chapter 6 – Latent Dirichlet Allocation (LDA)

6.1 BLOG TOPIC ANALYSIS USING LDA

Topic modeling focuses on the problem of classifying sets of documents into themes, either manually or automatically (preferred). It is a way of identifying patterns in a corpus-technically, this is text mining. We take our corpus and group words across the corpus into 'topics,' using a text mining algorithm or tool.

> ***Definition 6.1.*** *In natural language processing, Latent Dirichlet Allocation (LDA) is a generative statistical model that allows sets of observations to be explained by unobserved groups that explain why some parts of the data are similar.*

To tell briefly, LDA assumes a fixed set of topics, where each topic represents a set of words. The goal of LDA is to map all the documents to the topics in a manner, such that the words in each document are mostly captured by those assumed topics. The model infers word posterior probabilities in topics from joint distributions.

We will use a collection of 20 posts from my LinkedIn blog as an example corpus, call "blog_corpus." The corpus can be downloaded at *https://github.com/stricje1/VIT_University/tree/master/Data_Analytics_2018/data*.

```
#Load text mining library
library(tm)
```

```
Loading required package: NLP
```

6.1.1 *Set working Directory*

See your working direct and modify path as needed. Alternatively, we can move or copy the files to the current working directory.

```
setwd("C:/Users/jeff/Documents/VIT_Course_Material/Data_Analytic
s_2018/code/blog_corpus/")
```

Here, we get a listing of .txt files in the directory.

```
filenames <- list.files(getwd(),pattern="*.txt")
```

6.1.2 Read files into a character vector

After we set the directory where our files are located, we read them line-by-line into a vector of characters.

```
files <- lapply(filenames,readLines)
```

6.1.3 Create corpus from vector

Next, we create the corpus (see **Definition 2.1**) from the vector of characters.

```
docs <- Corpus(VectorSource(files))
# inspect a particular document in corpus
writeLines(as.character(docs[1]))
```

```
c("10 More Signs that you might be a Data Scientist", "Why are
people so down on Data Scientists? Probably because they don't k
now what one looks like. There are a lot of folks out there call
ing themselves Data Scientist just because they had some introdu
ctory statistic courses, and maybe they can make really cool cha
rts in Excel. All that makes you is someone who had some introdu
ctory statistic courses and can make some really cool charts in
Excel. So, what might a real Data Scientist look like? Here are
some more signs that you might \"really\" be a Data Scientist (s
ee 10 Signs that you might be a Data Scientist, October 7, 2014)
.",
"1.\tYou have a title of Operation Research Analyst, Principal
Business Analyst, Senior Risk Analyst, or Statistician, Data Arc
hitect, Senior Data Analyst, and so on.", "2.\tYou have a gradua
te degree or graduate course work in a related field, i.e., Anal
ytics, Predictive Modeling, Risk Analysis, Applied Statistics, D
atabase Architecture, etc.", "3.\tYou have a professionally take
n greyscale photo on your LinkedIn profile. Who does that?", "4.
\tIn addition to Data Science, you include some or all or the fo
llowing skills on your profile: SPSS Modeler, SAS Programming, R
egression Analysis, Statistical Modeling, Logistics Regression,
Machine Learning, SQL, Python, etc.",
"5.\tYour boss or customer listens to you and takes action when
you present your analysis and recommendations.", "6.\tNot only c
an you crunch numbers with the best of them, you can logically p
resent the results of your analysis in plain language, or at lea
st your customerâ\200\231s language.", "7.\tYou write incredibly
Geek-like post on LinkedIn, whether there are many viewers or no
t. Who does that?", "8.\tProfit was made or money was saved (or
for non-profits some measure of performance was achieved) based
on your work.",
"9.\tYou belong to a LinkedIn group for predictive modeling, pr
escriptive modeling, analytics, business analytics, etc.", "10.\
tOne of your hobbies is data analysis and modeling. Who would do
```

```
that?")
 list(language = "en")
 list()
```

6.1.4 Start Preprocessing

The next preprocessing steps include converting to lower case, removing special characters/symbols, removing punctuation, stripping digits, removing stop words, abd removing white spaces.

First, we transform the corpus to lower case.

```
docs <-tm_map(docs,content_transformer(tolower))
```

Second, we remove punctuation.

```
docs <- tm_map(docs, removePunctuation)
```

Third, we strip digits from the corpus.

```
docs <- tm_map(docs, removeNumbers)
```

Fourth, we remove stopwords.

```
docs <- tm_map(docs, removeWords, stopwords("english"))
```

Fitfh, we romove our own set of custom stopwords.

```
docs <- tm_map(docs, removeWords, myStopwords)
#define and eliminate all custom stopwords
myStopwords <- c("can", "say","one","way","use",
        "also","howev","tell","will",
        "much","need","take","tend","even",
        "like","particular","rather","said",
        "get","well","make","ask","come","end",
        "first","two","help","often","may",
        "might","see","someth","thing","point",
        "post","look","right","now","think",
        "anoth","put","set","new","good",
        "want","sure","kind","yes,","day","etc",
        "quit","sinc","attempt","lack","seen",
        "littl","ever","moreov","though","abl",
        "enough","earli","away","achiev","draw",
        "last","never","brief","entir","brief",
        "great","lot")
```

Finally, we remove whitespaces.

```
docs <- tm_map(docs, stripWhitespace)
It is good practice to check the document preprocessing every no
w and then.
writeLines(as.character(docs[2]))
```

```
 cten signs might data scientist nailed really good definition d
ata scientist argue need entity poorly defined perform one follo
wing task might data scientist tyou build databased models predi
ctive descriptive prescriptive tyou perform data mining analysis
tyou use data perform analytics business otherwise tyou can spel
l data scientist using binary numbers tyou regularly use sas ent
erprise guide sas enterprise miner spss modeler spss statistics
tyou can program r use perform data analysis modeling tyou reall
y hardcore use matlab octave data analysis tyou can spell data s
cientist latin ready little sas book tyou entire sas library ipa
d iphone etc tyou use data provide added value company customer
ten signs might data scientist tyou title operation research ana
lyst principal business analyst senior risk analyst statistician
tyou graduate degree graduate course work
 list(language = "en")
 list()
```

6.1.5 Stem document

If the blogs we were analyzing were written in Great Britian, there might be differences between our English and British english 2), we might want to account for differences like *analyze* and *analyse*.

```
docs <- tm_map(docs,stemDocument)
docs <- tm_map(docs, content_transformer(gsub),
               pattern = "organiz", replacement = "organ")
docs <- tm_map(docs, content_transformer(gsub),
               pattern = "organis", replacement = "organ")
docs <- tm_map(docs, content_transformer(gsub),
               pattern = "andgovern", replacement = "govern")
docs <- tm_map(docs, content_transformer(gsub),
               pattern ="inenterpris",replacement = "enterpris")
docs <- tm_map(docs, content_transformer(gsub),
               pattern = "team-", replacement = "team")
docs <- tm_map(docs, content_transformer(gsub),
               pattern = "henry", replacement = "henry")
```

At this point, it is a good idea to inspect a document as a check agian.

```
writeLines(as.character(docs[3]))
```

```
 c type analyt user sell thursday afternoon analyt user meet bec
om standard mayb youâr familiar similar meet key metric everyon
follow present present deck everi number slice dice bar chart pi
```

```
e chart dozen page  differ version process  whatâ realli happen
peopl  question most peopl just took  meet peopl thank present t
aken  time  number togeth hung accomplish  declar meet wast time
theyâr wast time analyt mean differ  differ peopl â" differ anal
yt user  donât understand peopl consum analyt wonât absorb  less
on critic analyt solut provid youâr sell differ user mean solut
adapt differ user messag  adapt analyt user user analyt break  d
imens purpos analyt versus skill  user   axe form box diagram a
cross bottom userâ purpos analyt either exploratori decisionmak
along side userâ technic skill level lower technic skill higher
technic skill lower technic skill user  someon understand metric
graph chart produc advanc analyt output higher technic skill use
r  understand general advanc statist output twodimens breakdown
creat four major type analyt user analyst scientist expert execu
t type analyt user break categori type analyt user differ  type
softwar  differ type support requir  differ analyst letâ start
explor analyst user analyt user statist savvi someon understand
databas queri simpli possess sophist advanc technic skill  stati
st model construct role primarili queri data number differ  unde
rstand relationship
 list(language = "en")
 list()
```

6.1.6 *Create document-term matrix*

The next code chunk generates the document-term matrix (see **Definition 2.5**) for our corpus, and the next several code chucks set up the document-term matrix for our analysis.

```
dtm <- DocumentTermMatrix(docs)
```

Then, we use this code chunk to convert rownames to filenames.

```
rownames(dtm) <- filenames
```

The following code chunk collapses the matrix by summing over columns

```
freq <- colSums(as.matrix(dtm))
```

This code chunk determines the length, which should be total number of terms.

```
length(freq)
```

```
 [1] 2833
```

The next code chunk creates the sort order (descending by frequency) for our terms and writes it to our disk.

```
ord <- order(freq,decreasing=TRUE)
write.csv(freq[ord],"word_freq.csv")
```

6.1.7 Load Topic Models Library

Here we load the *topicmodels* package that we will be using in our analysis.

```
library(topicmodels)
```

6.1.8 Set Parameters for Gibbs Sampling

With the following code chunk, we set the parameters for the LDA using Gibbs sampling that we will implement below. In statistics, Gibbs sampling or a Gibbs sampler is a *Markov Chain Monte Carlo* (MCMC) algorithm for obtaining a sequence of observations which are approximated from a specified multivariate probability distribution, when direct sampling is difficult. It is a simple and often highly effective approach for performing posterior inference in probabilistic models.

Definition 6.2. *Suppose that a sample X is taken from a distribution depending on a parameter vector $\theta \in \Theta$ of length d, with prior distribution $g(\theta_1, \dots, \theta_n)$. It may be that d is very large and that numerical integration to find the marginal densities of the θ_i would be computationally expensive. Then an alternative method of calculating the marginal densities is to create a Markov chain on the space Θ by repeating these two steps:*

1. Pick a random index $1 \leq j \leq d$

2. Pick a new value for θ_i according to $g(\theta_1, \dots \theta_{j-1}, \cdot, \theta_{j+1}, \dots, \theta_d)$

These steps define a **reversible Markov chain** *with the desired invariant distribution g.*

We also make a rough guess for the number of topics as five. Later, we will use a user function to make a better guess at the optimal number of topics, k.

```
burnin <- 4000
iter <- 2000
thin <- 500
seed <-list(2003,5,63,100001,765)
nstart <- 5
best <- TRUE
```

6.1.9 Run LDA using Gibbs sampling

Using the parameters, we set above, we now implement the LDA() model using Gibbs sampling. As a matter of convenience, we use the key word control to set the parameters.

```
library(topicmodels)
control=list(nstart=5, seed = list(2003,5,63,100001,765), best=T
RUE, burnin = 4000, iter = 2000, thin=500)
ldaOut <-LDA(dtm,5, method="Gibbs", control=control)
```

6.1.10 Write out Results

Now we write the documents to topics in a CSV file. This optional, but it could save us a lot of work if our system crashed, for instance.

```
ldaOut.topics <- as.matrix(topics(ldaOut))
write.csv(ldaOut.topics,file=paste("LDAGibbs",k,
        "DocsToTopics.csv"))
```

6.1.10.1 Top 6 Terms per Topic

The next code chunk provides the top six terms for each topic.

```
ldaOut.terms <- as.matrix(terms(ldaOut,6))
write.csv(ldaOut.terms,file=paste("LDAGibbs",k,
        "TopicsToTerms.csv"))
```

6.1.10.2 Topic Probabilities

Here, we calculate the probabilities associated with each topic assignment.

```
topicProbabilities <- as.data.frame(ldaOut@gamma)
write.csv(topicProbabilities,file=paste("LDAGibbs",k,
        "TopicProbabilities.csv"))
```

Then, we find relative importance of top 2 topics.

```
topic1ToTopic2 <- lapply(1:nrow(dtm),function(x)
sort(topicProbabilities[x,])[k]/sort(topicProbabilities[x,])
        [k-1])
```

Next, we find relative importance of second and third most important topics:

```
topic2ToTopic3 <- lapply(1:nrow(dtm),function(x)
sort(topicProbabilities[x,])[k-1]/sort(topicProbabilities[x,])
       [k-2])
```

Finally, we write the output to a file:

```
write.csv(topic1ToTopic2,file=paste("LDAGibbs",k,
      "Topic1ToTopic2.csv"))
write.csv(topic2ToTopic3,file=paste("LDAGibbs",k,
       "Topic2ToTopic3.csv"))
```

6.2 2012 USA PRESIDENTIAL DEBATE USING LDA

The bipartisan Commission on Presidential Debates (CPD) held four debates for the 2012 U.S. presidential general election, slated for various locations around the United States in October 2012 – three of them involving the major party presidential nominees (later determined to be Democratic President Barack Obama from Illinois and former Republican Governor Mitt Romney of Massachusetts), and one involving the vice-presidential nominees (Vice President Joe Biden from Delaware and Representative Paul Ryan of Wisconsin) (Little, 2012).

In this section we will perform a topic analysis of the debates from transcripts that are publically available. The debate transcripts favailableat*:*

https://github.com/pedrosan/TheAnalyticsEdge.

6.2.1 Loading and Installing Libraries

Here, we install and load libraries needed for data processing and plotting:

```
if (!require("pacman")) install.packages("pacman")
pacman::p_load_gh("trinker/gofastr")
pacman::p_load(tm, topicmodels, dplyr, tidyr, igraph, devtools,
LDAvis, ggplot2, sentimentr)
library(scales)
library(Rcpp)
library(tm)
library(topicmodels)
library(dplyr)
library(tidyr)
library(igraph)
library(devtools)
```

```
library(LDAvis)
library(ggplot2)
library(sentimentr)
```

6.2.2 Get External Scripts

Next, we call external script containing the `optimal_k()` function definition, using its source function call. (For the full optimal_k source code, see Section 6.3.). Optimal k uses the silhouette meathod that recursively measures how similar a point is to its own cluster (cohesion) compared to other clusters (separation).

```
invisible(lapply(
       file.path("https://raw.githubusercontent.com/trinker/topi
cmodels_learning/master/functions",
       c("topicmodels2LDAvis.R", "optimal_k.R")),
       devtools::source_url))
```

The content of this external file is included in the Appendix at the end of this document.

6.2.3 Load the Data

Now, we load the data, Presidential Debates 2012, and view the first five records.

```
data(presidential_debates_2012)
head(presidential_debates_2012,5)
```

```
 # A tibble: 5 x 5
   person tot   time   role    dialogue
   <fct>  <chr> <fct>  <fct>   <chr>
 1 LEHRER 1.1   time 1 modera~ We'll talk about specifically abo
ut health ~
 2 LEHRER 1.2   time 1 modera~ But what do you support the vouch
er system,~
 3 ROMNEY 2.1   time 1 candid~ What I support is no change for c
urrent ret~
 4 ROMNEY 2.2   time 1 candid~ And the president supports taking
dollar se~
 5 LEHRER 3.1   time 1 modera~ And what about the vouchers?
```

In this code chunk we generate user-defined stopwords that are relevant to our case.

```
stops <- c(
        tm::stopwords("english"),
        tm::stopwords("SMART"), "governor",
        "president", "mister", "obama","romney"
        ) %>%
        gofastr::prep_stopwords()
```

The next code chunk creates the document-term matrix.

```
doc_term_mat <- presidential_debates_2012 %>%
with(gofastr::q_dtm_stem(dialogue,
paste(person, time, sep = "_"))) %>%
gofastr::remove_stopwords(stops, stem=TRUE) %>%
gofastr::filter_tf_idf()
```

This code chunk established the control list we will use to find `optimum_k()`.

```
control <- list(burnin = 500, iter = 1000, keep = 100)
```

6.2.4 Determine Optimal Number of Topics

Using the external `optimal_k()` function code, we determine the optimal number of topics, which is a little better than a random guess.

```
k <- optimal_k(doc_term_mat, 40, control = control)
```

```
 Grab a cup of coffee this could take a while...
 10 of 40 iterations (Current: 05:32:19; Elapsed: .1 mins)
 20 of 40 iterations (Current: 05:32:30; Elapsed: .3 mins; Remai
ning: ~.7 mins)
 30 of 40 iterations (Current: 05:32:47; Elapsed: .5 mins; Remai
ning: ~.4 mins)
 40 of 40 iterations (Current: 05:33:12; Elapsed: 1 mins; Remain
ing: ~0 mins)
 Optimal number of topics = 17
```

Figure 6-1. Graph of document-term matrix with 20 nodes. This gives us the number of topics—the Harmonic Mean of Log Likelihood. The plot is created from the document-term matrix (see **Definition 2.5**).

```
print.optimal_k(doc_term_mat, 40, control = control)
```

```
[1] "A graph with 20 nodes."
```

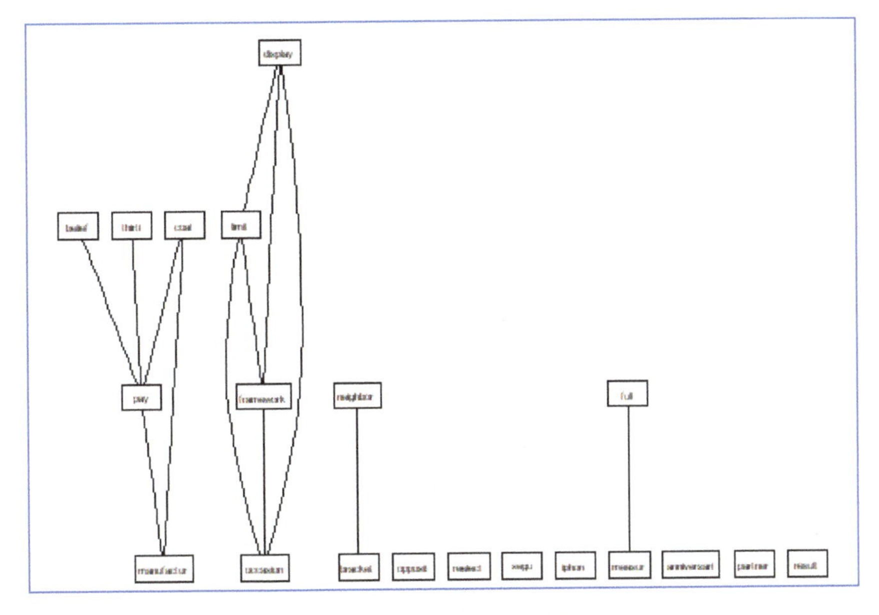

Figure 6-1. *Graph of document-term matrix with 20 nodes.*

6.2.5 Run the Model

Now, we implement he LDA model using Gibbs sampling (see **Definition 6.2**) and the parameters we have already established. An optional sampling method is the variational expectation-maximization (VEM) algorithm (Grun & Hornik, 2011).

```
control[["seed"]] <- 100
lda_model <- topicmodels::LDA(doc_term_mat, k=as.numeric(k),
        method = "Gibbs",
        control = control)
```

6.2.6 Plot the Topics Per Person & Time

This code chunk is used to generate a plot in Figure 6-2. Topics of Presidential debates of the topics per person and time. However, in this instance we are only concerned with candidates Obama, Romney, Lehrer, Schiffer, Crowley, and Question. This is implemented in the code with the `levels()` argument in the `mutate()` function. The model generates topics (and the words that comprise them) from each candidate over time (three different debates). Notice the frequencis of topic discussion per candidate. It would appear that Romney, for instance, focused mostly on one topic per debate at the expense of not concentrating on others.

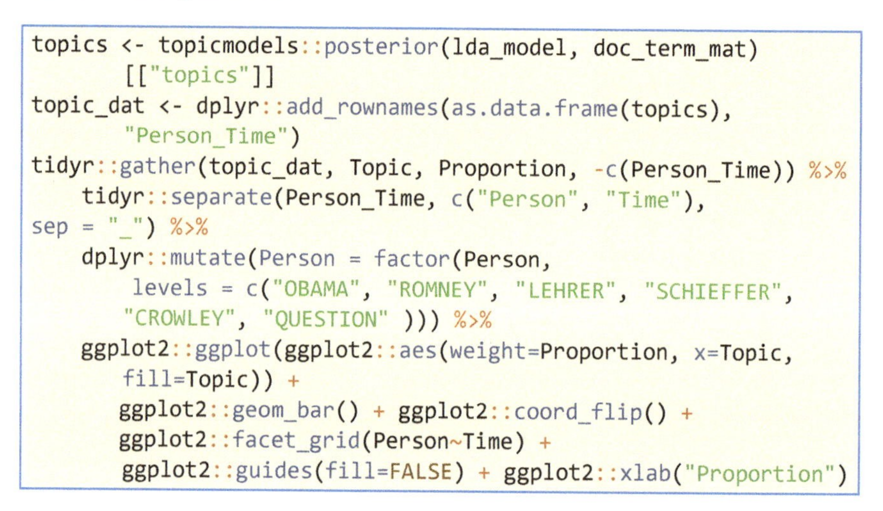

```
topics <- topicmodels::posterior(lda_model, doc_term_mat)
      [["topics"]]
topic_dat <- dplyr::add_rownames(as.data.frame(topics),
      "Person_Time")
tidyr::gather(topic_dat, Topic, Proportion, -c(Person_Time)) %>%
    tidyr::separate(Person_Time, c("Person", "Time"),
sep = "_") %>%
    dplyr::mutate(Person = factor(Person,
        levels = c("OBAMA", "ROMNEY", "LEHRER", "SCHIEFFER",
       "CROWLEY", "QUESTION" ))) %>%
    ggplot2::ggplot(ggplot2::aes(weight=Proportion, x=Topic,
       fill=Topic)) +
       ggplot2::geom_bar() + ggplot2::coord_flip() +
       ggplot2::facet_grid(Person~Time) +
       ggplot2::guides(fill=FALSE) + ggplot2::xlab("Proportion")
```

Figure 6-2. Topics of Presidential debates

The plot represents the topics of the 2012 Presidential debates between President Obama and Governor Romney. Jim Lehrer of PBS, Candy Crowley of CNN, and Bob Scheiffer of CBS were the moderators for the debates of October 3, October 16, and October 22, respectively. The plot shows the topics on the vertical axis and the count (proportion) on the horizaontal axis, the bars showing the proportion that each candidate (and moderators) discussed each topic.

6.2.7 *Plot the Topics Matrix as a Heatmap*

Now, we make a heatmap of our results, by candidate. The results are shown in Figure 6-3. The "hot" part of then map is red and indicates a high frequency of words to topics

```
heatmap(topics, scale = "none")
```

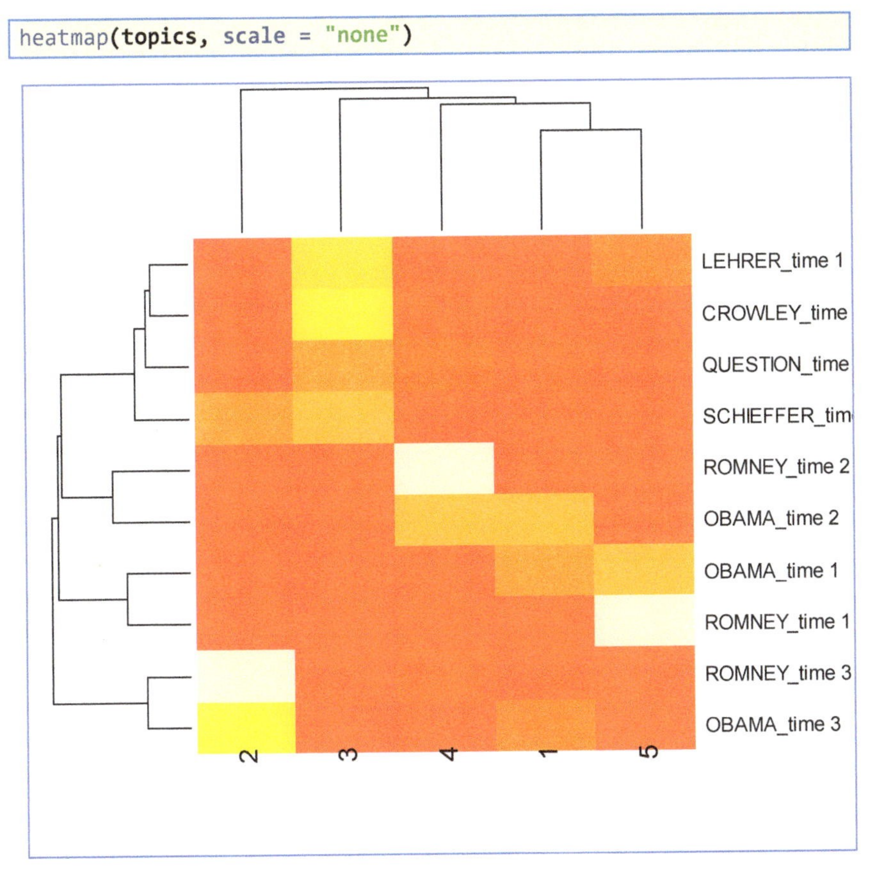

Figure 6-3. *Topics by Candidate*

6.2.8 Network of the Word Distributions Over Topics

Correlation in topic models can be considered in two forms: (1) the correlation in topic distributions, the correlation between topics; and (2) the correlation in topic-word distributions, the correlation between words.

Here, we perform the latter, and generate a network of word distributions over topics and we get the strength between topics based on word posterior probabilities, shown in Figure 6-4. The size of the circles represents the strength og the word probabilities to topics, with larger corresponding to stronger. For example, during the first debate, Romney did not utter the words that comprise topic 5, indicated by the "cool" color of the map. During the second debate, he did talk about topic 5 but did not say much about topic 4. The five topics come from the left axis of Figure 2.

```
post <- topicmodels::posterior(lda_model)
cor_mat <- cor(t(post[["terms"]]))
cor_mat[ cor_mat < .05 ] <- 0
diag(cor_mat) <- 0
graph <- graph.adjacency(cor_mat, weighted=TRUE, mode="lower")
graph <- delete.edges(graph, E(graph)[ weight < 0.05])

E(graph)$edge.width <- E(graph)$weight*20
V(graph)$label <- paste("Topic", V(graph))
V(graph)$size <- colSums(post[["topics"]]) * 15

par(mar=c(0, 0, 3, 0))
plot.igraph(graph, edge.width = E(graph)$edge.width,
    edge.color = "orange", vertex.color = "orange",
    vertex.frame.color = NA, vertex.label.color = "grey30")
title("Strength Between Topics Based On Word Probabilities", cex
.main=.8)
```

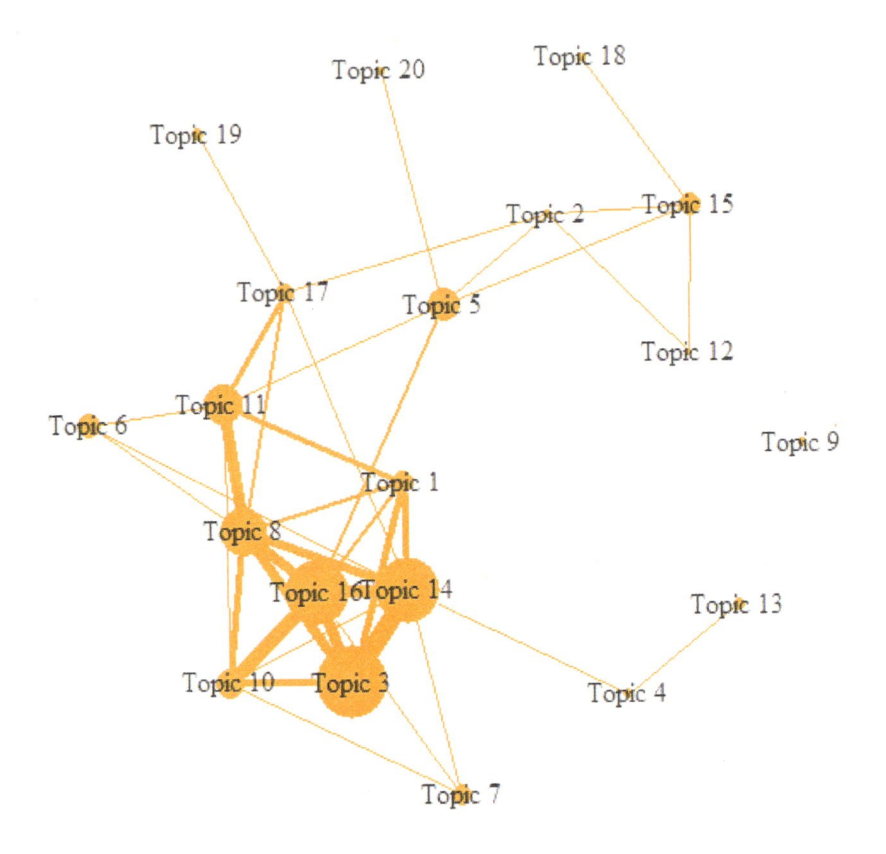

Figure 6-4. *Strength between topics based on word probabilities.*

6.2.9 Topic Distributions over Candidates

This code chunk generates a network of the topics over documents (candidates), with different looking graph shown in Figure 6-5. The concept of exploring topic distribution over documents is similar o what we have already explored. The graph is based on the incidence matrix. The topic to document incidence matrix is derived from the the LDA model with Gibbs sampling for determining the posterior probability distribution of topics to documents, which is how we define topic_mat in the second line of code. The third line of code generates the graph from the incidence matix data.

```
minval <- .1
topic_mat <- topicmodels::posterior(lda_model)[["topics"]]

graph <- graph_from_incidence_matrix(topic_mat, weighted=TRUE)
graph <- delete.edges(graph, E(graph)[ weight < minval])

E(graph)$edge.width <- E(graph)$weight*17
E(graph)$color <- "blue"
V(graph)$color <- ifelse(grepl("^\\d+$", V(graph)$name),
   "grey75", "orange")
V(graph)$frame.color <- NA
V(graph)$label <- ifelse(grepl("^\\d+$", V(graph)$name), paste
   ("topic", V(graph)$name), gsub("_", "\n",   V(graph)$name))
V(graph)$size <- c(rep(10, nrow(topic_mat)),
   colSums(topic_mat) * 20)
V(graph)$label.color <- ifelse(grepl("^\\d+$", V(graph)$name),
   "red", "grey30")

par(mar=c(0, 0, 3, 0))
set.seed(365)
plot.igraph(graph, edge.width = E(graph)$edge.width,
    vertex.color = adjustcolor(V(graph)$color, alpha.f = .4))
title("Topic & Document Relationships", cex.main=.8)
```

Notice the time axes in Figure 6-5 denoting the three time periods for the debates. Then observe that we can determine that both Governor Romley and Senator Obama participated in all three debates shown. Also, notice that Topic 8 was a key topic of discussion, by the size of the circle.

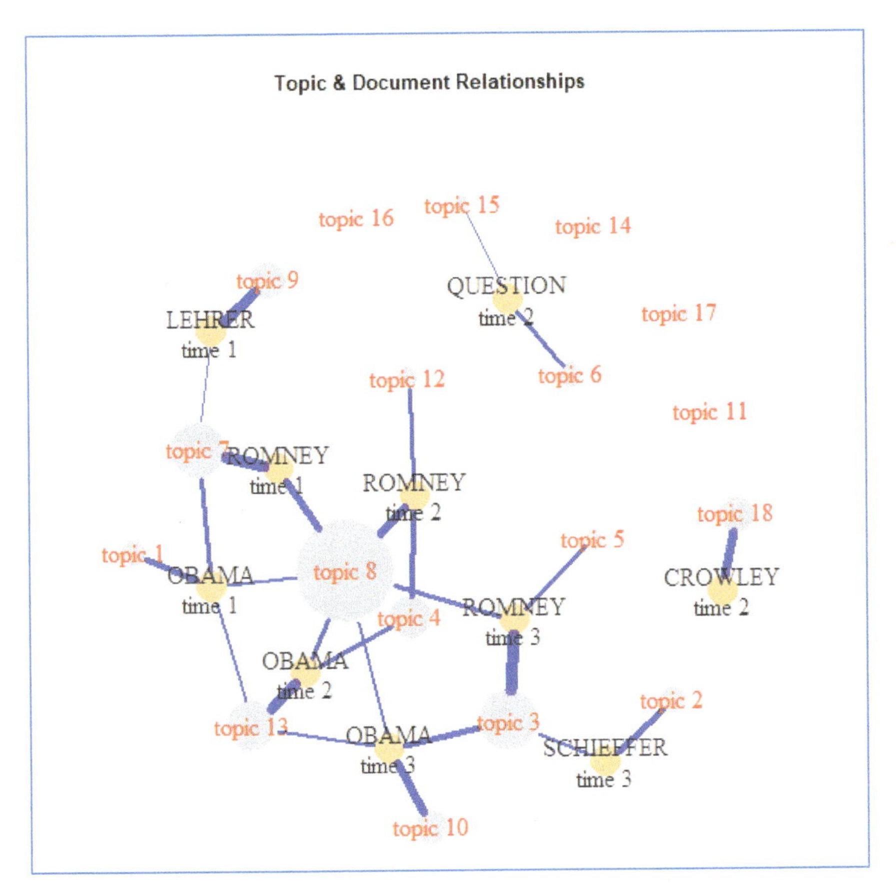

Figure 6-5. *Relationship between topics and documents.*

The topic and document relationships provide us with Topics by Candidate (document).

6.2.10 LDAvis of Model

The LDAvis provides an interactive view of the results by topic number, as shown in Figure 6-6. We call the function from an external script. Note that topic number 11 has been selected on the html dashboard.

```
source("D:\\Documents\\VIT_Course_Material\\Data_Analytics_2018\
\code\\topicmodels2LDAvis.R")
lda_model %>%
    topicmodels2LDAvis() %>%
    LDAvis::serVis()
```

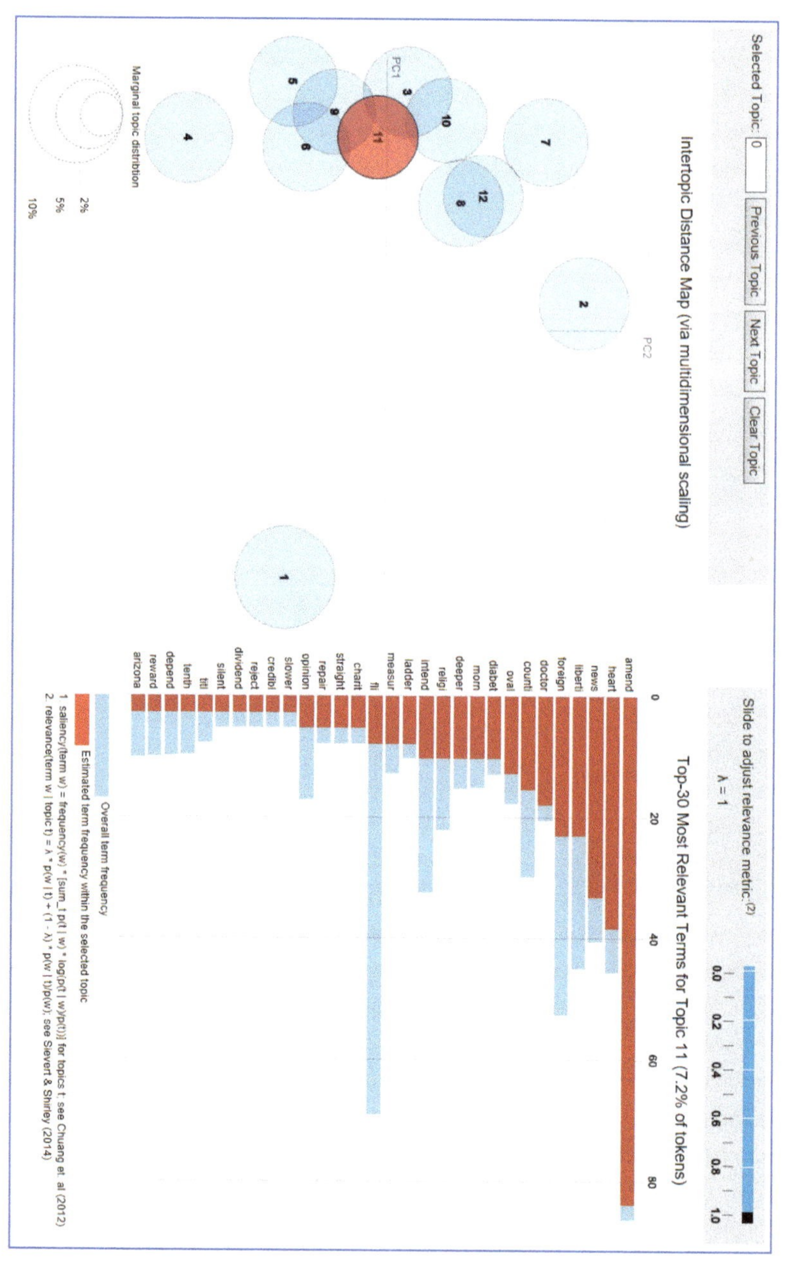

Figure 6-6. *Interactive view of the results by topic number.*

The function in the external script can be written and executed in your R script. The code is:

```
topicmodels2LDAvis <- function(x, ...){
  post <- topicmodels::posterior(x)
  if (ncol(post[["topics"]]) < 3) stop("The model must contain >
     2 topics")
  mat <- x@wordassignments
  LDAvis::createJSON(
     phi = post[["terms"]],
     theta = post[["topics"]],
     vocab = colnames(post[["terms"]]),
     doc.length = slam::row_sums(mat, na.rm = TRUE),
     term.frequency = slam::col_sums(mat, na.rm = TRUE)
  )
}
```

6.2.11 Build a New Model

Next, we load a new data set named partial_republican_debates_2015, which is a smaller data set. The next chunk of code pulls in the new data set from the *gofastr* package.

```
# Fitting New Data
library(gofastr)
# Create the DocumentTermMatrix for New Data
doc_term_mat2 <- partial_republican_debates_2015 %>%
    with(gofastr::q_dtm_stem(dialogue, paste(person, location,
        sep = "_"))) %>%
    gofastr::remove_stopwords(stops, stem=TRUE) %>%
    gofastr::filter_tf_idf() %>%
    gofastr::filter_documents()
```

Next, we run optimal_k again and in addition, we compute the harmonic means of log likelihood and show its graph in Figure 6-7. This gives us the optimal number of k-topics, which is 18, after the optimization (optimal_k) is stopped (at the red circle).

```
control = list(nstart=5, seed = list(2003,5,63,100001,765),
       best=TRUE, burnin = 4000, iter = 2000, thin=500)
control2  = list(burnin = 500, iter = 1000, keep = 100)
#control2[["estimate.beta"]] <- FALSE
k <- optimal_k1(doc_term_mat2,40,control=control2)
```

```
Grab a cup of coffee this could take a while...
10 of 40 iterations (Current: 05:33:21; Elapsed: .1 mins)
20 of 40 iterations (Current: 05:33:32; Elapsed: .3 mins; Remai
ning: ~.7 mins)
```

```
 30 of 40 iterations (Current: 05:33:50; Elapsed: .6 mins; Remai
ning: ~.4 mins)
 40 of 40 iterations (Current: 05:34:15; Elapsed: 1 mins; Remain
ing: ~0 mins)
 Optimal number of topics = 16
plot.optimal_k1(optimal_k1(doc_term_mat2,40,control=control2))

 Grab a cup of coffee this could take a while...
 10 of 40 iterations (Current: 05:34:23; Elapsed: .1 mins)
 20 of 40 iterations (Current: 05:34:34; Elapsed: .3 mins; Remai
ning: ~.7 mins)
 30 of 40 iterations (Current: 05:34:52; Elapsed: .6 mins; Remai
ning: ~.4 mins)
 40 of 40 iterations (Current: 05:35:18; Elapsed: 1 mins; Remain
ing: ~0 mins)
 Optimal number of topics = 18
```

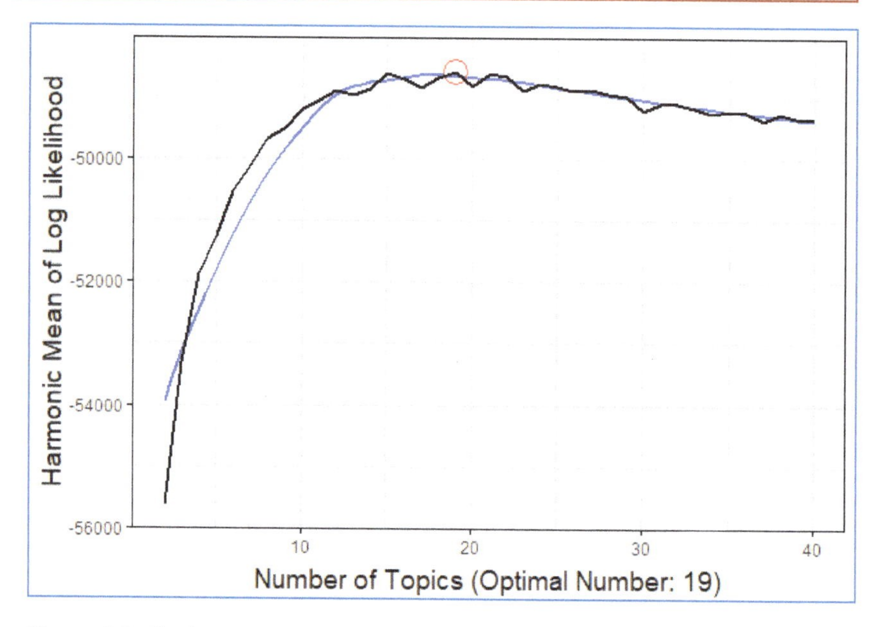

Figure 6-7. *The harmonic means of log likelihood for optimal topics.*

We also use function from an external script. We have it saved as a separate R file and called using `source()`, but the code can be called from your R script by executing it before requiring its use. The code follows:

```
topicmodels2LDAvis <- function(x, ...){
  post <- topicmodels::posterior(x)
  if (ncol(post[["topics"]]) < 3) stop("The model must contain >
     2 topics")
```

```
  mat <- x@wordassignments
  LDAvis::createJSON(
    phi = post[["terms"]],
    theta = post[["topics"]],
    vocab = colnames(post[["terms"]]),
    doc.length = slam::row_sums(mat, na.rm = TRUE),
    term.frequency = slam::col_sums(mat, na.rm = TRUE)
  )
}
```

The code is initialized by:

```
lda_model %>%
    topicmodels2LDAvis() %>%
    LDAvis::serVis()
library(gofastr)
## Create the DocumentTermMatrix for New Data
data(partial_republican_debates_2015)
doc_term_mat <- partial_republican_debates_2015 %>%
    with(gofastr::q_dtm_stem(dialogue, paste(person, location,
        sep = "_"))) %>%
    gofastr::remove_stopwords(stops, stem=TRUE) %>%
    gofastr::filter_tf_idf() %>%
    gofastr::filter_documents()
```

Next, we construct the model for the new debate data.

```
lda_model2 <- topicmodels::LDA(doc_term_mat2, k = as.numeric(k),
    model = lda_model, control = control2)
```

6.2.12 Plot the Topics Per Person & Location for New Data

In Figure 6-8 we plot the topics per person and location of the new data. You should notice that the left axis is topics based on words, the top axis is locations, the right axis is candidates, and the botoom axis is frequency. The aggregated colomn (NA) shows topics distribution for all candidates.

```
topics2 <- topicmodels::posterior(lda_model2, doc_term_mat2)
      [["topics"]]
topic_dat2 <- dplyr::add_rownames(as.data.frame(topics2),
      "Person_Location")
colnames(topic_dat2)[-1] <- apply(terms(lda_model2, 10), 2,
      paste, collapse = ", ")
tidyr::gather(topic_dat2, Topic, Proportion, -c(Person_Location)
      ) %>%
   tidyr::separate(Person_Location, c("Person", "Location"),
      sep = "_")
```

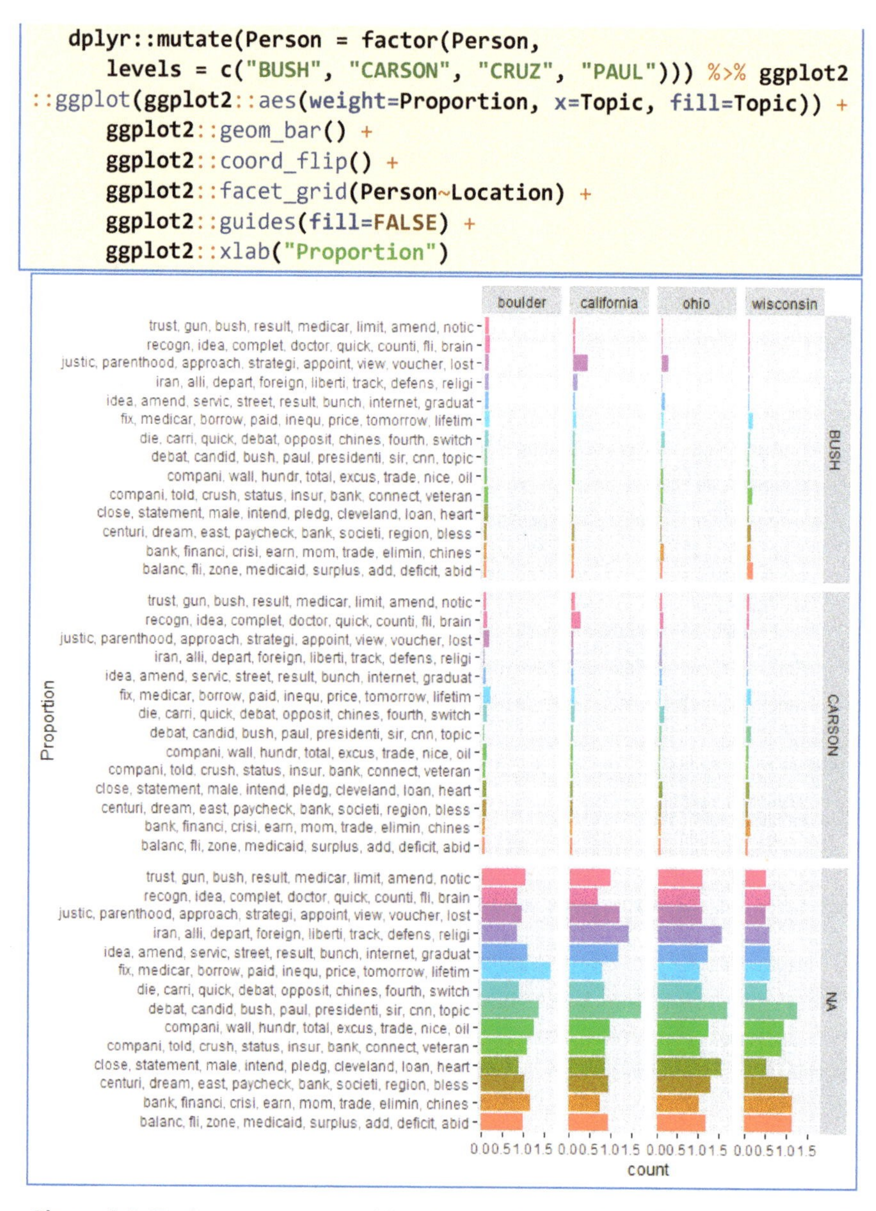

```
    dplyr::mutate(Person = factor(Person,
       levels = c("BUSH", "CARSON", "CRUZ", "PAUL"))) %>% ggplot2
::ggplot(ggplot2::aes(weight=Proportion, x=Topic, fill=Topic)) +
       ggplot2::geom_bar() +
       ggplot2::coord_flip() +
       ggplot2::facet_grid(Person~Location) +
       ggplot2::guides(fill=FALSE) +
       ggplot2::xlab("Proportion")
```


Figure 6-8. Topics per person and location of the new data. This is a portion of the entire graph, which is tabloid-size. Column N is the aggregated count.

Error! Reference source not found. is the heatmap for the preceding hierarchical model. It makes the interpretation more intuitive and is easier to read. The darker colors are "hotter" than the lighter one. For instance, the plot show that Trapper really honed in on Topic 13 in California.

```
heatmap(topics2, scale = "none")
```

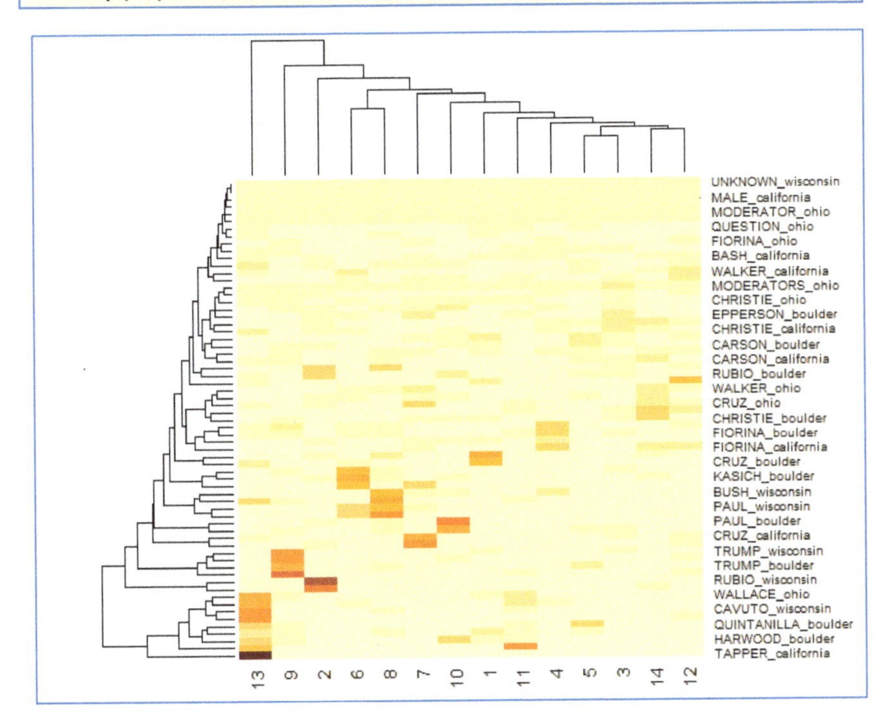

Figure 6-9. *Heatmap for the preceding hierarchical model, with candidates and location (person_location)*

6.3 Optimal_k

Optimal_k finds the optimal number of K-means clusters. The source code can be downloaded or called from:

https://rdrr.io/github/mcallaghan/scimetrix/src/R/optimal_k.R

6.4 Exercises

For the purpose of this exercise, we will be using the collection of Harry Potter novels. This package holds the collection of Harry Potter books 1-7 by J.K. Rowling. Install these using:

```
if (packageVersion("devtools") < 1.6) {
install.packages("devtools")
}
devtools::install_github("bradleyboehmke/harrypotter")
library(harrypotter)
```

We will also be using the following packages:

```
library(topicmodels) # topic modeling functions
library(stringr)     # common string functions
library(tidytext)    # tidy text analysis
library(tidyverse)   # data manipulation and visualization
library(scales)      # used for percent scale on confusion table
```

1. Perform preprocessing tasks
   a. Create a tibble of the document, which is a combined column of title and chapter, and a text column which contains the corpus. [Hint: The harrypotter package makes indexing the text by chapter easy, as each book is already separated into its chapters.]
   b. Use unnest_tokens to take each individual word from the corpus.
   c. Remove stop_words using anti_join from the `tidytext` package.
   d. Remove any proper name and other proper nouns that not add value to the meaning we are trying to extract from our data [Hint: look at the top words using top_n(word_counts, 10).
   e. Transform our term frequencies into a document-term matrix (`dtm`).
2. Perform Latent Dirichlet Allocation (LDA)
   a. Use to find the mixture of words that make up each topic and the mixture of topics that make up each document.
   b. Plot the top terms

c. how much each document is associated with each topic.

d. Inspect visually how well our unsupervised learning was able to distinguish between the topics for each of the titles.

3. Unsupervised Classification to assign topics

   a. Use `augment` function to add the count of each term next to the topic the term has been assigned to. [Hint: This allows us to know how much each word weighs in on the topic assignment of the chapter, and ultimately title, as a whole.]

   b. Investigate the proportion of assignments from one titles were assigned to another title using the confusion table.

   c. Which terms were the were most frequent in the mis-assigned topics?

R Glossary

R-Studio contain one of the best help tools we have seen. When you want to know what function to use for an operation, you type in a key word or phrase, and if you know the function you want to use, it will show you how it is used and the parameters it requires. However, it does nothing if you do not have a general idea of the functionality you need. Hence, we provide a list o ffunctions used in this text, with a brief explanation. Also, the index contains a list of R packages used and the pages where they are implemented.

R function	Description	Package
acast	cast a molten data frame into a vector/matrix/array	reshape2
add_rownames	convert row names to an explicit variable	dplyr
anti_join	anti_join that prints information about the operation	tidylog
rbind	combine R Objects by Rows	
as.Date	date Conversion Functions to and from character	base
as.matrix	convert a data.table to a matrix	data.table
bind_tf_idf	bind the term frequency and inverse document frequency of a tidy text dataset to the dataset	tidytext
cbind	combine R Objects by Columns	base
colSums	form Row and Column Sums and Means	base
comparison.cloud	plot a comparison cloud	wordcloud
coord_flip	cartesian coordinates with x and y flipped	ggplot2
Corpus	representing and computing on corpora.	tm
corpus	construct a corpus object	quanteda
data.frame	used as the fundamental data structure by most of R's modeling software	base
dcast	cast a molten data frame into a data frame	
dim	retrieve or set the dimension of an object.	base

DocumentTermMatrix	constructs or coerces to a term-document matrix or a document-term matrix	tm
expand_limits	sometimes you may want to ensure limits include a single value, for all panels or all plots.	ggplot2
facet_wrap	wraps a 1d sequence of panels into 2d, which is generally a better use of screen space than facet_grid() because most displays are roughly rectangular	ggplot2
file.path	construct the path to a file from components in a platform-independent way	base
filter	picks cases based on their values	dplyr
findAssocs	find associations in a document-term or term-document matrix	tm
findFreqTerms	find frequent terms in a document-term or term-document matrix	tm
function	these functions provide the base mechanisms for defining new functions in the R language	base
gather	gather takes multiple columns and collapses into key-value pairs, duplicating all other columns as needed	tidyr
geom_abline	add reference lines to a plot, either horizontal, vertical, or diagonal; these are useful for annotating plots	ggplot2
geom_bar	makes the height of the bar proportional to the number of cases in each group	ggplot2
geom_histogram	display the counts with bars; frequency polygons	ggplot2
geom_jitter	adds a small amount of random variation to the location of each point, and is a useful way of handling overplotting caused by discreteness in smaller datasets	ggplot2
geom_line	connects observations in order of the variable on the x axis.	ggplot2
geom_text	adds only text to the plot	ggplot2
get_sentiments	get specific sentiment lexicons in a tidy format	tidytext

getwd	returns an absolute filepath representing the current working directory of the R process	base
ggplot	initializes a ggplot object	tiddyverse
ggraph	this function is the equivalent ofggplot2::ggplot()in ggplot2; takes care of setting up the plotobject along with creating the layout for the plot	ggraph
glm	is used to fit generalized linear models, specified by giving a symbolic description of the linear predictor	stats
graph.adjacency	is a flexible function for creating igraph graphs from adjacency matrices	igraph
group_by	takes an existing tbl and converts it into a grouped tbl where operations are performed "by group"	dplyr
gsub	replaces all matches of a string	base
head	returns the first or last parts of a vector, matrix, table, data frame or function	utils
heatmap	this is an Axes-level function and will draw the heatmap into the currently-active Axes if none is provided to the ax argument	seaborn
inner_join	return all rows from x where there are matching values in y, and all columns from x and y	dplyr
inspect	display detailed information on a corpus, a term-document matrix, or a text document	tm
install.packages	download and install packages from CRAN-like repositories or from local files	utils
join	join two tbls together	dplyr
labs	used to the axis and legend labels display the full variable name.	ggplot2
lapply	returns a list of the same length as X	base
LDA	estimate a LDA model using for example the VEM algorithm or Gibbs Sampling	topicmodel s
lemmatize_strings	lemmatize a vector of strings.	textstem
lemmatize_words	lemmatize a vector of words.	textstem
length	get or set the length of vectors (including lists) and factors, and of any	base

	other R object for which a method has been defined	
library	library and require load and attach add-on packages	base
list	functions to construct, coerce and check for both kinds of R lists	base
mutate	adds new variables that are functions of existing variables	dplyr
nest	creates a list of data frames containing all the nested variables: this seems to be the most useful form in practice	tidyr
order	returns a permutation which rearranges its first argument into ascending or descending order, breaking ties by further arguments	base
pacman::p_load	this function is a wrapper for library and require. It checks to see if a package is installed, if not it	pacman
paste0	concatenate vectors after converting to character.	infix
pie3D	displays a 3D pie chart with optional labels	plotrix
posterior	calculate posterior for one item given score, difficulty and prior	dscore
prep_stowords	join multiple vectors of words, convert to lower case, and return sorted unique words	tm
read.csv	reads a file in comma separated format and creates a data frame from it, with cases corresponding to lines and variables to fields in the file	utils
read.delim	reads a file in tab delimited format and creates a data frame from it, with cases corresponding to lines and variables to fields in the file	utils
read.table	reads a file in table format and creates a data frame from it, with cases corresponding to lines and variables to fields in the file	utils
readHTMLTable	this function and its methods provide somewhat robust methods for extracting data from HTML tables in an HTML document	XML

remove_stopwords	remove stopwords and < nchar words from a Term Document Matrix or Document Term Matrix	tm
removeSparseTerms	remove sparse terms from a document-term or term-document matrix	tm
require	load and attach add-on packages	base
rownames	these functions allow to you detect if a data frame has row names (has_rownames()), remove them (remove_rownames()), or convert them back-and-forth between an explicit column (rownames_to_column() and column_to_rownames())	tibble
scale_continuous	scale_x_continuous() and scale_y_continuous() are the default scales for continuous x and y aesthetics	ggplot2
select	picks variables based on their names	dplyr
setwd	is used to set the working directory	base
sort	Sort (or order) a vector or factor (partially) into ascending or descending order	base
spread	spread a key-value pair across multiple columns	tidyr
str	compactly display the internal structure of an R object, a diagnostic function and an alternative to summary (and to some extent, dput)	utils
str_detect	detect the presence or absence of a pattern in a string	stringr
str_sub	extract and replace substrings from a character vector	stringr
str_to_lower	convert case of a string to lowercase	stringr
summarize	reduces multiple values down to a single summary	dplyr
summary	summary is a generic function used to produce result summaries of the results of various model fitting functions.	base
TermDocumentMatrix	constructs or coerces to a term-document matrix or a document-term matrix.	tm

theme	themes are a powerful way to customize the non-data components of your plots: i.e. titles, labels, fonts, etc.	ggplot2
tibble	onstructs a data frame. It is used like base::data.frame(), but with a couple notable differences	tibble
tm_map	interface to apply transformation functions (also denoted as mappings) to corpora	tm
tokenize_words	perform basic tokenization into words	tokenizers
tokenize_sentenc es	perform basic tokenization into sentences	tokenizers
tokenize_charact ers	perform basic tokenization into characters	tokenizers
tokens	construct a tokens object, either by importing a named list of characters from an external tokenizer, or by calling the internal quanteda tokenizer	quanteda
tokens_select	select tokens with only positive pattern matches from a list of regular expressions, including a dictionary	quanteda
tokens_wordstem	this is a wrapper to wordStem designed to allow this function to be called without loading the entire SnowballC package	topicmodel s
top_n	convenient wrapper that uses filter() and min_rank() to select the top or bottom entries in each group	dplyr
trimws	remove leading and/or trailing whitespace from character strings	base
ungroup	removes grouping	dplyr
unnest_tokens	Split a column into tokens using the tokenizers package, splitting the table into one-token-per-row	tidytext
wordcloud2	Function for Creating wordcloud by wordcloud	wordcloud2
write.csv	write to csv format	utils
writeLines	write text lines to a connection	base
tally	counts/tally observations by group	dplyr
count	count the number of occurences	plyr

inner_join	return all rows from x where there are matching values in y, and all columns from x and y	dplyr
tokenize	simple regular expression-based parser that splits the components of a vector of character into tokens while protecting infix punctuation	tau

Python Glossary

The Python glosaary is "lighter" than the R glossary. In Pyton, there are fewer functions and libraries required to do the same amount of work in R. Also, the online help for NLTK, Scikit-Learm, Numpy, Pandas, and Matplotlib is excellent.

Python function	Description	Library
accuracy_score	In multilabel classification, this function computes subset accuracy	sklearn.metrics
classification_report	Build a text report showing the main classification metrics.	sklearn.metrics
confusion_matrix	Compute confusion matrix to evaluate the accuracy of a classification	sklearn.metrics
CountVectorizer	Convert a collection of text documents to a matrix of token counts	sklearn.feature_ extraction.text.
f1_score	sklearn.linear_model	sklearn.metrics
LinearSVC()	Linear Support Vector Classification.	sklearn.svm
LogisticRegression	Logistic Regression (aka logit, MaxEnt) classifier	sklearn. linear_model
metrics.recall_score	The recall is intuitively the ability of the classifier to find all the positive samples	sklearn.metrics
MultinomialNB	Naive Bayes classifier for multinomial models	sklearn. naive_bayes
nltk.word_tokenize	Return a tokenized copy of text, using NLTK's recommended word tokenizer	NLTK
numpy.array	Create an array	Numpy
numpy.asarray	Convert the input to an array	Numpy
numpy.loadtxt	Load data from a text file	Numpy
open	Open and return file-like object	Numpy
pd.DataFrame	Two-dimensional, size-mutable, potentially heterogeneous tabular data	Pandas
pd.read_csv	Read a comma-separated values (csv) file into DataFrame	Pandas
Perceptron	Perceptron is a classification algorithm which shares the same	sklearn. linear_model

	underlying implementation with SGDClassifier	
PorterStemmer	remove morphological affixes from words, leaving only the word stem.	NLTK
precision_score	The precision is intuitively the ability of the classifier not to label as positive a sample that is negative.	sklearn.metrics
pyplot.bar	Make a bar plot	Pyplot
pyplot.figure	Create a new figure, or activate an existing figure.	Pyplot
pyplot.grid	Configure the grid lines.	Pyplot
pyplot.hist	Plot a histogram.	Pyplot
pyplot.imshow	Display data as an image, i.e., on a 2D regular raster.	Pyplot
pyplot.legend	Place a legend on the axes	Pyplot
pyplot.subplots	Add a subplot to the current figure.	Pyplot
pyplot.title	Set one of the three available axes titles	Pyplot
pyplot.xlabel	Set the label for the x-axis.	Pyplot
pyplot.xticks	Get or set the current tick locations and labels of the x-axis	Pyplot
pyplot.ylabel	Set the label for the y-axis	Pyplot
pyplot.yticks	Get or set the current tick locations and labels of the y-axis	Pyplot
pypolt.show	Display a pyplot object	Pyplot
RandomForestClassifier	A random forest is a meta estimator that fits a number of decision tree classifiers	sklearn. ensemble
RidgeClassifier	This classifier first converts the target values into {-1, 1} and then treats problem as a regression task	sklearn. linear_model
SGDClassifier	This estimator implements regularized linear models with stochastic gradient descent (SGD) learning	sklearn. linear_model
sns.distplot	Flexibly plot a univariate distribution of observations	Seaborn
sns.heatmap	Plot rectangular data as a color-encoded matrix	Seaborn

sns.stripplot	Draw a scatterplot where one variable is categorical	sklearn.feature_ extraction
TfidfTransformer	Transform a count matrix to a normalized tf or tf-idf representation	sklearn.feature_ extraction.text.
TfidfVectorizer	Convert a collection of raw documents to a matrix of TF-IDF features	sklearn.feature_ extraction.text.
train_test_split	Split arrays or matrices into random train and test subsets	sklearn. model_selection
word_tokenize	Tokenizers divide strings into lists of substrings (words)	NLTK
WordCloud	Displays a list of words, the importance of each beeing shown with font size or color	Wordcloud
WordNetLemmatizer	Returns the input word unchanged if it cannot be found in WordNet	NLTK

REFERENCES

Americanbar.org. (2018, November 1). *Servicemembers Civil Relief Act (SCRA)*. Retrieved from American Bar Association: https://www.americanbar.org/groups/legal_services/milvets/aba_home_front/information_center/servicemembers_civil_relief_act/

BBC. (2016, July 26). France church attack: Priest killed by two 'IS militants'. *BBC News Online*. Retrieved from https://www.bbc.com/news/world-europe-36892785

Bing, L., & Minqing, H. (2004). Mining and summarizing customer reviews. *Proceedings of the ACM SIGKDD International Conference on Knowledge Discovery & Data Mining (KDD-2004).* Seattle, Washington. Retrieved from https://www.cs.uic.edu/~liub/FBS/sentiment-analysis.html

Breiman, L., H., F. J., Olshen, R., & Stone, C. (1984). *Classification and Regression.* Monterey, CA.: Wadsworth and Brooks. Retrieved from ftp://ftp.stat.berkeley.edu/pub/users/breiman/OOBestimation.ps.Z.

CFPB. (2020, January 2020). *Consumer Complaint Data Catalog.* Retrieved from Consumer Financial Protection Bureau: https://catalog.data.gov/dataset/consumer-complaint-database

Data Society. (2016). How ISIS Uses Twitter. *Data World*, Online. Retrieved from https://data.world/data-society/how-isis-uses-twitter

Expert Systems Team. (2016, April 11). *Natural language processing and text mining*. Retrieved from Expert Systems: Basic Semantics Blog: https://expertsystem.com/natural-language-processing-and-text-mining/

Feinerer, I. (2019). *Introduction to the tm Package.* 12: December. Retrieved from https://rdrr.io/rforge/tm/f/inst/doc/tm.pdf

Great Learning. (2020, April 29). Machine Learning Interview Questions and Answer for 2020 you must prepare. *Great Learning Blog,* Online. Retrieved from https://www.mygreatlearning.com/blog/machine-learning-interview-questions/

Grun, B., & Hornik, K. (2011). Topicmodels: An R Package for Fitting Topic Models. *Journal of Statistical Software*, 1-20.

Kaemingk, B., Rembado, M. C., Shroff, M., Entwistle, J., & Aiello, S. (2016, February 8). Building a Unique Cultural Prediction Engine for the Academy Awards. *Measuring the Digital World*, Online. Retrieved from http://measuringthedigitalworld.com/tag/concept-mapping/

Kass, G. V. (1980). An exploratory technique for investigating large quantities of categorical data. *Applied Statistics, 29*(2), 119–127. doi:10.2307/2986296

Kenton, W. (2019, April 17). *Regulation E*. Retrieved from Investopedia: https://www.investopedia.com/terms/r/regulation-e.asp

Kenton, W. (2019, September 9). *Regulation Z*. Retrieved from Investopedia: https://www.investopedia.com/terms/r/regulation_z.asp

Lexico.com. (2020). *Lexico Dictionary*. (D. a. Press, Editor) Retrieved January 16, 2020, from Lexico: https://www.lexico.com/definition/natural_language

Little, M. (2012, June 25). Presidential debate formats announced, feature town hall. *Los Angeles Times*, p. Online.

Loughran, T., & McDonald, B. (2016, April 26). Textual Analysis in Accounting and Finance: A Survey. *Journal of Accounting Research, 54*(3), 1187-1230. doi:https://doi.org/10.1111/1475-679X.12123

Mikolov, T., Le, Q. V., & Sutskever, I. (2013, September 17). Exploiting Similarities among Languages for Machine Translation. Retrieved from http://citeseerx.ist.psu.edu/viewdoc/summary?doi=10.1.1.754.2995

Milward, D. (2020). *What is Text Mining, Text Analytics and Natural Language Processing?* Retrieved from Linguamatics: https://www.linguamatics.com/what-text-mining-text-analytics-and-natural-language-processing

Prinzie, A., & Van den Poel, D. (2008). Random Forests for multiclass classification: Random MultiNomial Logit". *Expert Systems with Applications, 34*(3), 1721–1732.

Project Jupyter. (2020). The Jupyter Notebook. *Project Jupyter.* Retrieved from https://jupyter.org/

Rite, S. (2018, December 15). *Demystifying 'Confusion Matrix' Confusion.* Retrieved from Toward Data Science: https://towardsdatascience.com/demystifying-confusion-matrix-confusion-9e82201592fd

Segal, T. (2019, July 5). Prescriptive Analytics. *Investopedia,* Online. Retrieved from https://www.investopedia.com/terms/p/prescriptive-analytics.asp

SHRM. (2015, July 27). *Complying with and Leveraging the Affordable Care Act* . Retrieved from Society for Human Resource Management (SHRM): https://www.shrm.org/resourcesandtools/tools-and-samples/toolkits/pages/complyingwithandleveragingtheaffordablecareact.aspx

Staff. (2020, July 1). *PYPL PopularitY of Programming Language.* Retrieved from PYPL: http://pypl.github.io/PYPL.html

Tolosi, L., & Lengauer, T. (2011). Classification with correlated features: unreliability of feature ranking and solutions. *Bioinformatics.* doi:10.1093/bioinformatics/btr300

Willems, K. (2018, November 20). *This R Data Import Tutorial Is Everything You Need.* Retrieved from Datacamp.com: https://www.datacamp.com/community/tutorials/r-data-import-tutorial#txt

Woolman, M. (2006). *Ways of Knowing: Introduction to the Theory of Knowledge.* IBID Press.

WPR Staff. (2020). *Colorado Springs, Colorado Population 2020.* Retrieved from World Population Review: https://worldpopulationreview.com/us-cities/colorado-springs-population/

INDEX

A

B

C

D

E

F

G

H

I

J

K

L

M

N

O

P

S

T

V

W

Z